SpringerBriefs in Mathematics

SpringerBriefs in Mathematics showcases expositions in all areas of mathematics and applied mathematics. Manuscripts presenting new results or a single new result in a classical field, new field, or an emerging topic, applications, or bridges between new results and already published works, are encouraged. The series is intended for mathematicians and applied mathematicians.

More information about this series at http://www.springer.com/series/10030

SpringerBriefs present concise summaries of cutting-edge research and practical applications across a wide spectrum of fields. Featuring compact volumes of 50 to 125 pages, the series covers a range of content from professional to academic. Briefs are characterized by fast, global electronic dissemination, standard publishing contracts, standardized manuscript preparation and formatting guidelines, and expedited production schedules.

Typical topics might include:

- A timely report of state-of-the art techniques
- A bridge between new research results, as published in journal articles, and a contextual literature review
- A snapshot of a hot or emerging topic
- An in-depth case study
- A presentation of core concepts that students must understand in order to make independent contributions

Titles from this series are indexed by Web of Science, Mathematical Reviews, and zbMATH.

Silvestru Sever Dragomir

Kato's Type Inequalities for Bounded Linear Operators in Hilbert Spaces

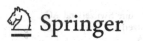

Silvestru Sever Dragomir
Department of Mathematics, College
of Engineering and Science
Victoria University
Melbourne, VIC, Australia

School of Computer Science and Applied
Mathematics, DST-NRF Centre
of Excellence in the Mathematical
and Statistical Sciences
University of the Witwatersrand
Johannesburg, South Africa

ISSN 2191-8198   ISSN 2191-8201   (electronic)
SpringerBriefs in Mathematics
ISBN 978-3-030-17458-3   ISBN 978-3-030-17459-0   (eBook)
https://doi.org/10.1007/978-3-030-17459-0

Mathematics Subject Classification (2010): 47A63, 47A50, 47A99

This Springer imprint is published by the registered company Springer Nature Switzerland AG.
The registered company address is: Gewerbestrasse 11, 6330 Cham, Switzerland

This book is dedicated to my granddaughters Sienna and Audrey.

Preface

Linear operator theory in Hilbert spaces plays a central role in contemporary mathematics with numerous applications for partial differential equations, in approximation theory, optimization theory, numerical analysis, probability theory and statistics and other fields.

The main aim of this book is to present several results related to Kato's famous inequality for bounded linear operators on complex Hilbert spaces obtained by the author in a sequence of recent research papers.

The book is intended for use by both researchers in various fields of linear operator theory and mathematical inequalities, domains which have grown exponentially in the last decade, as well as by postgraduate students and scientists applying inequalities in their specific areas.

The monograph starts with a short introductory chapter where the famous Kato's inequality is introduced and some simple, however, important particular cases are revealed.

In the second chapter, we present several multiplicative and additive generalizations of Kato's inequality for n-tuple of bounded linear operators on the Hilbert space $(H; \langle \cdot, \cdot \rangle)$. Applications for functions of normal operators defined by power series and inequalities for *Euclidian norm*, *s-1-norm* and *s-1-numerical radius* of n-tuples of operators are provided as well.

In the third chapter, we present a two-parameter generalization of Kato due to Furuta. Applications for functions of bounded linear operators defined by power series, inequalities for four bounded operators generalizing Furuta's inequality and some general *norm* and *numerical radius* inequalities are given as well.

In the fourth chapter, after recalling some fundamental facts on *Hilbert–Schmidt operators, trace operators* and some properties of traces of such operators, we present a trace version of Kato's inequality. Some natural functionals associated with this inequality and their superadditivity and monotonicity are established. Several inequalities for sequences of operators and power series of operators are given as well.

In the fifth chapter, after recalling some fundamental facts on Bochner integral for measurable functions with values in Banach spaces, we provide an integral version of Kato's inequality. Several *norm* and *numerical radius* inequalities with applications for the *operator exponential* are also given.

For the sake of completeness, all the results presented are completely proved and the original references where they have been firstly obtained are mentioned.

Melbourne, Australia   Silvestru Sever Dragomir

Contents

Chapter 1
Introduction

We denote by $\mathcal{B}(H)$ the Banach algebra of all bounded linear operators on a complex Hilbert space $(H; \langle \cdot, \cdot \rangle)$.

If P is a positive selfadjoint operator on H, i.e. $\langle Px, x \rangle \geq 0$ for any $x \in H$, then the following inequality is a generalization of the Schwarz inequality in H

$$|\langle Px, y \rangle|^2 \leq \langle Px, x \rangle \langle Py, y \rangle, \tag{1.1}$$

for any $x, y \in H$.

The following inequality is of interest as well, see [21, p. 221].

Let P be a positive selfadjoint operator on H. Then

$$\|Px\|^2 \leq \|P\| \langle Px, x \rangle \tag{1.2}$$

for any $x \in H$.

The *"square root"* of a positive bounded selfadjoint operator on H can be defined as follows, see for instance [21, p. 240]: *If the operator $A \in B(H)$ is selfadjoint and positive, then there exists a unique positive selfadjoint operator $B := \sqrt{A} \in B(H)$ such that $B^2 = A$. If A is invertible, then so is B.*

If $A \in \mathcal{B}(H)$, then the operator A^*A is selfadjoint and positive. Define the *"absolute value"* operator by $|A| := \sqrt{A^*A}$.

In 1952, Kato [22] proved the following celebrated generalization of Schwarz inequality for any bounded linear operator T on H:

$$|\langle Tx, y \rangle|^2 \leq \langle (T^*T)^\alpha x, x \rangle \langle (TT^*)^{1-\alpha} y, y \rangle, \tag{1.3}$$

for any $x, y \in H$, $\alpha \in [0, 1]$. Utilizing the modulus notation introduced before, we can write (1.3) as follows

© The Author(s), under exclusive license to Springer Nature Switzerland AG 2019
S. S. Dragomir, *Kato's Type Inequalities for Bounded Linear Operators in Hilbert Spaces*, SpringerBriefs in Mathematics,
https://doi.org/10.1007/978-3-030-17459-0_1

$$\left|\langle Tx, y\rangle\right|^2 \le \left\langle |T|^{2\alpha} x, x\right\rangle \left\langle \left|T^*\right|^{2(1-\alpha)} y, y\right\rangle \tag{1.4}$$

for any $x, y \in H$, $\alpha \in [0, 1]$.

It is useful to observe that, if $T = N$, a normal operator, i.e., we recall that $NN^* = N^*N$, then the inequality (1.4) can be written as

$$\left|\langle Nx, y\rangle\right|^2 \le \left\langle |N|^{2\alpha} x, x\right\rangle \left\langle |N|^{2(1-\alpha)} y, y\right\rangle, \tag{1.5}$$

and in particular, for selfadjoint operators A we can state it as

$$\left|\langle Ax, y\rangle\right| \le \left\| |A|^\alpha x \right\| \left\| |A|^{1-\alpha} y \right\| \tag{1.6}$$

for any $x, y \in H$, $\alpha \in [0, 1]$.

If $T = U$, a unitary operator, i.e., we recall that $UU^* = U^*U = 1_H$, then the inequality (1.4) becomes

$$\left|\langle Ux, y\rangle\right| \le \|x\| \, \|y\|$$

for any $x, y \in H$, which provides a natural generalization for the Schwarz inequality in H.

The symmetric powers in the inequalities above are natural to be considered, so if we choose in (1.4), (1.5) and in (1.6) $\alpha = 1/2$ then we get for any $x, y \in H$

$$\left|\langle Tx, y\rangle\right|^2 \le \langle |T| x, x\rangle \left\langle \left|T^*\right| y, y\right\rangle, \tag{1.7}$$

$$\left|\langle Nx, y\rangle\right|^2 \le \langle |N| x, x\rangle \langle |N| y, y\rangle, \tag{1.8}$$

and

$$\left|\langle Ax, y\rangle\right| \le \left\| |A|^{1/2} x \right\| \left\| |A|^{1/2} y \right\| \tag{1.9}$$

respectively.

It is also worthwhile to observe that, if we take the supremum over $y \in H$, $\|y\| = 1$ in (1.4) then we get

$$\|Tx\|^2 \le \|T\|^{2(1-\alpha)} \left\langle |T|^{2\alpha} x, x\right\rangle \tag{1.10}$$

for any $x \in H$, or in an equivalent form

$$\|Tx\| \le \left\| |T|^\alpha x \right\| \|T\|^{1-\alpha} \tag{1.11}$$

for any $x \in H$.

If we take $\alpha = 1/2$ in (1.10), then we get

$$\|Tx\|^2 \le \|T\| \langle |T| x, x\rangle \tag{1.12}$$

for any $x \in H$, which in the particular case of $T = P$, a positive operator, provides the result from (1.2).

For various interesting generalizations, extension and Kato related results, see the papers [11, 19, 28, 29, 33].

In this monograph we present several recent inequalities related to Kato's famous result (1.3) obtained by the author in the sequence of research papers [3, 10].

Chapter 2
Inequalities for n-Tuples of Operators

In this chapter we present several multiplicative and additive generalizations of Kato's inequality for n-tuple of bounded linear operators on the Hilbert space $(H; \langle \cdot, \cdot \rangle)$. Applications for functions of normal operators defined by power series and inequalities for *Euclidian Norm, s-1-Norm* and *s-1-Numerical Radius* of n-tuples of operators are provided as well.

2.1 Multiplicative Inequalities

The following vector inequality holds:

Theorem 2.1 (Dragomir [3]) *Let* $(T_1, ..., T_n) \in \mathcal{B}(H) \times ... \times \mathcal{B}(H) := \mathcal{B}^{(n)}(H)$ *be an n-tuple of bounded linear operators on the Hilbert space* $(H; \langle \cdot, \cdot \rangle)$ *and* $(p_1, ..., p_n) \in \mathbb{R}^{*n}_+$ *an n-tuple of nonnegative weights not all of them equal to zero. Then we have*

$$\sum_{j=1}^{n} p_j \left| \langle T_j x, y \rangle \right|^2 \leq \left\langle \sum_{j=1}^{n} p_j \left| T_j \right|^2 x, x \right\rangle^{\alpha} \left\langle \sum_{j=1}^{n} p_j \left| T_j^* \right|^2 y, y \right\rangle^{1-\alpha} \tag{2.1}$$

for any $x, y \in H$ *with* $\|x\| = \|y\| = 1$ *and* $\alpha \in [0, 1]$.

Proof We must prove the inequalities only in the case $\alpha \in (0, 1)$, since the case $\alpha = 0$ or $\alpha = 1$ follows directly from the corresponding case of Kato's inequality.

Utilizing Kato's inequality for the operator T_j, $j \in \{1, ..., n\}$ we have

© The Author(s), under exclusive license to Springer Nature Switzerland AG 2019
S. S. Dragomir, *Kato's Type Inequalities for Bounded Linear Operators in Hilbert Spaces*, SpringerBriefs in Mathematics,
https://doi.org/10.1007/978-3-030-17459-0_2

$$\sum_{j=1}^{n} p_j \left|\langle T_j x, y\rangle\right|^2 \leq \sum_{j=1}^{n} p_j \left\langle \left|T_j\right|^{2\alpha} x, x\right\rangle \left\langle \left|T_j^*\right|^{2(1-\alpha)} y, y\right\rangle \qquad (2.2)$$

$$\leq \sum_{j=1}^{n} p_j \left\langle \left|T_j\right|^2 x, x\right\rangle^{\alpha} \left\langle \left|T_j^*\right|^2 y, y\right\rangle^{1-\alpha}$$

for any $x, y \in H$ with $\|x\| = \|y\| = 1$, where for the last inequality we have used the Hölder-McCarthy inequality $\langle P^r x, x\rangle \leq \langle Px, x\rangle^r$ that holds for any positive operator P and any power $r \in (0, 1)$.

Now, on making use of the weighted Hölder discrete inequality

$$\sum_{j=1}^{n} p_j a_j b_j \leq \left(\sum_{j=1}^{n} p_j a_j^p\right)^{1/p} \left(\sum_{j=1}^{n} p_j b_j^q\right)^{1/q}, \ p, q > 1, \frac{1}{p} + \frac{1}{q} = 1,$$

where $(a_1, ..., a_n), (b_1, ..., b_n) \in \mathbb{R}_+^n$, and choose $a_j = \left\langle \left|T_j\right|^2 x, x\right\rangle^{\alpha}, b_j = \left\langle \left|T_j^*\right|^2 y, y\right\rangle^{1-\alpha}, p = \frac{1}{\alpha}$ and $q = \frac{1}{1-\alpha}$ then we get

$$\sum_{j=1}^{n} p_j \left\langle \left|T_j\right|^2 x, x\right\rangle^{\alpha} \left\langle \left|T_j^*\right|^2 y, y\right\rangle^{1-\alpha} \qquad (2.3)$$

$$\leq \left\{\sum_{j=1}^{n} p_j \left[\left\langle \left|T_j\right|^2 x, x\right\rangle^{\alpha}\right]^{1/\alpha}\right\}^{\alpha} \left\{\sum_{j=1}^{n} p_j \left[\left\langle \left|T_j^*\right|^2 y, y\right\rangle^{1-\alpha}\right]^{1/(1-\alpha)}\right\}^{1-\alpha}$$

$$= \left\{\sum_{j=1}^{n} p_j \left\langle \left|T_j\right|^2 x, x\right\rangle\right\}^{\alpha} \left\{\sum_{j=1}^{n} p_j \left\langle \left|T_j^*\right|^2 y, y\right\rangle\right\}^{1-\alpha}$$

$$= \left\langle \sum_{j=1}^{n} p_j \left|T_j\right|^2 x, x\right\rangle^{\alpha} \left\langle \sum_{j=1}^{n} p_j \left|T_j^*\right|^2 y, y\right\rangle^{1-\alpha}$$

for any $x, y \in H$ with $\|x\| = \|y\| = 1$.

Utilizing (2.2) and (2.3) we deduce the desired inequality (2.1). ∎

Remark 2.2 The inequality (2.1) becomes for $y = x$ the following simpler result that is useful for deriving numerical radius inequalities:

$$\sum_{j=1}^{n} p_j \left|\langle T_j x, x\rangle\right|^2 \le \left\langle \sum_{j=1}^{n} p_j \left|T_j\right|^2 x, x \right\rangle^{\alpha} \left\langle \sum_{j=1}^{n} p_j \left|T_j^*\right|^2 x, x \right\rangle^{1-\alpha} \tag{2.4}$$

$$\le \left\langle \sum_{j=1}^{n} p_j \left[\alpha \left|T_j\right|^2 + (1-\alpha) \left|T_j^*\right|^2 \right] x, x \right\rangle$$

for any $x \in H$ with $\|x\| = 1$.

Let $(N_1, ..., N_n) \in \mathcal{B}^{(n)}(H)$ be an n-tuple of normal operators on the Hilbert space $(H; \langle \cdot, \cdot \rangle)$. Then from the above Theorem 2.1 we have the following result that can be utilized in obtaining various inequalities for functions of normal operators defined by power series, namely:

$$\sum_{j=1}^{n} p_j \left|\langle N_j x, y\rangle\right|^2 \le \left\langle \sum_{j=1}^{n} p_j \left|N_j\right|^2 x, x \right\rangle^{\alpha} \left\langle \sum_{j=1}^{n} p_j \left|N_j\right|^2 y, y \right\rangle^{1-\alpha} \tag{2.5}$$

for any $x, y \in H$ with $\|x\| = \|y\| = 1$, $\alpha \in [0, 1]$ and any n-tuple weights $(p_1, ..., p_n) \in \mathbb{R}_+^{*n}$.

In particular, we get from (2.5) the following inequality for modulus of normal operators

$$\sum_{j=1}^{n} p_j \left|\langle N_j x, x\rangle\right|^2 \le \left\langle \sum_{j=1}^{n} p_j \left|N_j\right|^2 x, x \right\rangle \tag{2.6}$$

for any $x \in H$ with $\|x\| = 1$.

The following result provides upper bounds for the sum $\sum_{j=1}^{n} p_j \left|\langle T_j x, y\rangle\right|$ and has important consequences in refining the fundamental triangle inequality for operator norm.

Theorem 2.3 (Dragomir [3]) *With the assumptions in Theorem 2.1 we have*

$$\sum_{j=1}^{n} p_j \left|\langle T_j x, y\rangle\right| \le \left\langle \sum_{j=1}^{n} p_j \left|T_j\right|^{2\alpha} x, x \right\rangle^{1/2} \left\langle \sum_{j=1}^{n} p_j \left|T_j^*\right|^{2(1-\alpha)} y, y \right\rangle^{1/2} \tag{2.7}$$

for any $x, y \in H$.

Proof From Kato's inequality for the operator T_j, $j \in \{1, ..., n\}$ we have

$$\sum_{j=1}^{n} p_j \left|\langle T_j x, y\rangle\right| \le \sum_{j=1}^{n} p_j \left\langle \left|T_j\right|^{2\alpha} x, x \right\rangle^{1/2} \left\langle \left|T_j^*\right|^{2(1-\alpha)} y, y \right\rangle^{1/2} \tag{2.8}$$

for any $x, y \in H$.

Now, on making use of the weighted Cauchy–Bunyakovsky–Schwarz discrete inequality

$$\sum_{j=1}^{n} p_j a_j b_j \leq \left(\sum_{j=1}^{n} p_j a_j^2\right)^{1/2} \left(\sum_{j=1}^{n} p_j b_j^2\right)^{1/2}$$

where $(a_1, ..., a_n), (b_1, ..., b_n) \in \mathbb{R}_+^n$, and choose $a_j = \left\langle |T_j|^{2\alpha} x, x\right\rangle^{1/2}$ and $b_j = \left\langle |T_j^*|^{2(1-\alpha)} y, y\right\rangle^{1/2}$, then we get

$$\sum_{j=1}^{n} p_j \left\langle |T_j|^{2\alpha} x, x\right\rangle^{1/2} \left\langle |T_j^*|^{2(1-\alpha)} y, y\right\rangle^{1/2} \tag{2.9}$$

$$\leq \left\{\sum_{j=1}^{n} p_j \left[\left\langle |T_j|^{2\alpha} x, x\right\rangle^{1/2}\right]^2\right\}^{1/2} \left\{\sum_{j=1}^{n} p_j \left[\left\langle |T_j^*|^{2(1-\alpha)} y, y\right\rangle^{1/2}\right]^2\right\}^{1/2}$$

$$= \left\{\sum_{j=1}^{n} p_j \left\langle |T_j|^{2\alpha} x, x\right\rangle\right\}^{1/2} \left\{\sum_{j=1}^{n} p_j \left\langle |T_j^*|^{2(1-\alpha)} y, y\right\rangle\right\}^{1/2}$$

$$= \left\langle \sum_{j=1}^{n} p_j |T_j|^{2\alpha} x, x\right\rangle^{1/2} \left\langle \sum_{j=1}^{n} p_j |T_j^*|^{2(1-\alpha)} y, y\right\rangle^{1/2}$$

for any $x, y \in H$.   ■

Remark 2.4 One of possible vector-valued extensions of (2.7) is as follows:

$$\sum_{j=1}^{n} p_j \left|\langle T_j x, x\rangle\right| \leq \left\langle \sum_{j=1}^{n} p_j |T_j|^{2\alpha} x, x\right\rangle^{1/2} \left\langle \sum_{j=1}^{n} p_j |T_j^*|^{2(1-\alpha)} x, x\right\rangle^{1/2} \tag{2.10}$$

$$\leq \left\langle \sum_{j=1}^{n} p_j \left[\frac{|T_j|^{2\alpha} + |T_j^*|^{2(1-\alpha)}}{2}\right] x, x\right\rangle$$

for any $x \in H$.

Remark 2.5 The symmetric case for powers, namely the case $\alpha = \frac{1}{2}$ in (2.7) is of interest since will produce the simpler result

$$\sum_{j=1}^{n} p_j \left|\langle T_j x, y\rangle\right| \leq \left\langle \sum_{j=1}^{n} p_j |T_j| x, x\right\rangle^{1/2} \left\langle \sum_{j=1}^{n} p_j |T_j^*| y, y\right\rangle^{1/2} \tag{2.11}$$

for any $x, y \in H$.

In particular, from (2.10) we derive

$$\sum_{j=1}^{n} p_j \left| \langle T_j x, x \rangle \right| \leq \left\langle \sum_{j=1}^{n} p_j \left| T_j \right| x, x \right\rangle^{1/2} \left\langle \sum_{j=1}^{n} p_j \left| T_j^* \right| x, x \right\rangle^{1/2} \qquad (2.12)$$

$$\leq \left\langle \sum_{j=1}^{n} p_j \left[\frac{\left| T_j \right| + \left| T_j^* \right|}{2} \right] x, x \right\rangle$$

for any $x \in H$.

Let $(N_1, \ldots, N_n) \in \mathcal{B}^{(n)}(H)$ be an n-tuple of normal operators on the Hilbert space $(H; \langle \cdot, \cdot \rangle)$. Then from the above Theorem 2.3 we have

$$\sum_{j=1}^{n} p_j \left| \langle N_j x, y \rangle \right| \leq \left\langle \sum_{j=1}^{n} p_j \left| N_j \right|^{2\alpha} x, x \right\rangle^{1/2} \left\langle \sum_{j=1}^{n} p_j \left| N_j \right|^{2(1-\alpha)} y, y \right\rangle^{1/2} \qquad (2.13)$$

for any $x, y \in H$.

In particular, we have

$$\sum_{j=1}^{n} p_j \left| \langle N_j x, x \rangle \right| \leq \left\langle \sum_{j=1}^{n} p_j \left| N_j \right|^{2\alpha} x, x \right\rangle^{1/2} \left\langle \sum_{j=1}^{n} p_j \left| N_j \right|^{2(1-\alpha)} x, x \right\rangle^{1/2} \qquad (2.14)$$

$$\leq \left\langle \sum_{j=1}^{n} p_j \left[\frac{\left| N_j \right|^{2\alpha} + \left| N_j \right|^{2(1-\alpha)}}{2} \right] x, x \right\rangle$$

for any $x \in H$.

2.2 Functional Inequalities

Now, by the help of power series $f(z) = \sum_{n=0}^{\infty} a_n z^n$ we can naturally construct another power series which will have as coefficients the absolute values of the coefficient of the original series, namely, $f_A(z) := \sum_{n=0}^{\infty} |a_n| z^n$. It is obvious that this new power series will have the same radius of convergence as the original series. We also notice that if all coefficients $a_n \geq 0$, then $f_A = f$, see also [32, p. 246].

As some natural examples that are useful for applications, we can point out that, if

$$f(z) = \sum_{n=1}^{\infty} \frac{(-1)^n}{n} z^n = \ln \frac{1}{1+z}, \quad z \in D(0, 1); \qquad (2.15)$$

$$g(z) = \sum_{n=0}^{\infty} \frac{(-1)^n}{(2n)!} z^{2n} = \cos z, \ z \in \mathbb{C};$$

$$h(z) = \sum_{n=0}^{\infty} \frac{(-1)^n}{(2n+1)!} z^{2n+1} = \sin z, \ z \in \mathbb{C};$$

$$l(z) = \sum_{n=0}^{\infty} (-1)^n z^n = \frac{1}{1+z}, \ z \in D(0,1);$$

then the corresponding functions constructed by the use of the absolute values of the coefficients are

$$f_A(z) = \sum_{n=1}^{\infty} \frac{1}{n} z^n = \ln \frac{1}{1-z}, \ z \in D(0,1); \tag{2.16}$$

$$g_A(z) = \sum_{n=0}^{\infty} \frac{1}{(2n)!} z^{2n} = \cosh z, \ z \in \mathbb{C};$$

$$h_A(z) = \sum_{n=0}^{\infty} \frac{1}{(2n+1)!} z^{2n+1} = \sinh z, \ z \in \mathbb{C};$$

$$l_A(z) = \sum_{n=0}^{\infty} z^n = \frac{1}{1-z}, \ z \in D(0,1).$$

The following result is a functional generalization of Kato's inequality for normal operators from (1.5).

Theorem 2.6 (Dragomir [3]) *Let $f(z) = \sum_{n=0}^{\infty} a_n z^n$ be a function defined by power series with real coefficients and convergent on the open disk $D(0,R) := \{z \in \mathbb{C}, \ |z| < R\}, \ R > 0$. If N is a normal operator on the Hilbert space H and for $\alpha \in (0,1)$ we have that $\|N\|^{2\alpha}, \|N\|^{2(1-\alpha)} < R$, then we have the inequalities*

$$|\langle f(N)x, y\rangle| \le \langle f_A (|N|^{2\alpha}) x, x\rangle^{1/2} \langle f_A (|N|^{2(1-\alpha)}) y, y\rangle^{1/2} \tag{2.17}$$

for any $x, y \in H$.
  In particular, if $\|N\| < R$, then

$$|\langle f(N)x, y\rangle| \le \langle f_A(|N|)x, x\rangle^{1/2} \langle f_A(|N|)y, y\rangle^{1/2} \tag{2.18}$$

for any $x, y \in H$.

Proof If N is a normal operator, then for any $j \in \mathbb{N}$ we have that

$$\left|N^j\right|^2 = (N^*N)^j = |N|^{2j}.$$

Now, utilising the inequality (2.13) we can write that

$$\left| \left\langle \sum_{j=0}^{n} a_j N^j x, y \right\rangle \right| \qquad (2.19)$$

$$\leq \sum_{j=0}^{n} |a_j| \left| \langle N^j x, y \rangle \right|$$

$$\leq \left\langle \sum_{j=0}^{n} |a_j| \left| N^j \right|^{2\alpha} x, x \right\rangle^{1/2} \left\langle \sum_{j=0}^{n} |a_j| \left| N^j \right|^{2(1-\alpha)} y, y \right\rangle^{1/2}$$

$$= \left\langle \sum_{j=0}^{n} |a_j| \left(|N|^{2\alpha} \right)^j x, x \right\rangle^{1/2} \left\langle \sum_{j=0}^{n} |a_j| \left(|N|^{2(1-\alpha)} \right)^j y, y \right\rangle^{1/2}$$

for any $x, y \in H$ and $n \in \mathbb{N}$.

Since $\|N\|^{2\alpha}, \|N\|^{2(1-\alpha)} < R$, then it follows that the series $\sum_{j=0}^{\infty} |a_j| \left(|N|^{2\alpha} \right)^j$ and $\sum_{j=0}^{\infty} |a_j| \left(|N|^{2(1-\alpha)} \right)^j$ are absolute convergent in $\mathcal{B}(H)$, and by taking the limit over $n \to \infty$ in (2.19) we deduce the desired result (2.17). ∎

Remark 2.7 Assume that f, R, N and α are as in Theorem 2.6. If we take the supremum in (2.17) over $y \in H$, $\|y\| = 1$, then we get

$$\|f(N) x\| \leq \left\langle f_A \left(|N|^{2\alpha} \right) x, x \right\rangle^{1/2} \left\| f_A \left(|N|^{2(1-\alpha)} \right) \right\|^{1/2} \qquad (2.20)$$

for any $x \in H$, which produces the operator norm inequality

$$\|f(N)\| \leq \left\| f_A \left(|N|^{2\alpha} \right) \right\|^{1/2} \left\| f_A \left(|N|^{2(1-\alpha)} \right) \right\|^{1/2}. \qquad (2.21)$$

If we take $y = x$ in (2.17), then we get

$$|\langle f(N) x, x \rangle| \leq \left\langle f_A \left(|N|^{2\alpha} \right) x, x \right\rangle^{1/2} \left\langle f_A \left(|N|^{2(1-\alpha)} \right) x, x \right\rangle^{1/2} \qquad (2.22)$$

$$\leq \left\langle \left[\frac{f_A \left(|N|^{2\alpha} \right) + f_A \left(|N|^{2(1-\alpha)} \right)}{2} \right] x, x \right\rangle$$

for any $x \in H$. This produces the following inequalities for the numerical radius

$$w(f(N)) \leq \begin{cases} \left\| f_A \left(|N|^{2\alpha} \right) \right\|^{1/2} \left\| f_A \left(|N|^{2(1-\alpha)} \right) \right\|^{1/2}; \\ \left\| \frac{f_A(|N|^{2\alpha}) + f_A(|N|^{2(1-\alpha)})}{2} \right\|. \end{cases} \qquad (2.23)$$

Making use of the examples in (2.15) and (2.16) we can state the vector inequalities:

$$\left|\langle \ln \left(1_H + N\right)^{-1} x, y\rangle\right|$$
$$\leq \langle \ln \left(1_H - |N|^{2\alpha}\right)^{-1} x, x\rangle^{1/2} \langle \ln \left(1_H - |N|^{2\alpha}\right)^{-1} y, y\rangle^{1/2}, \|N\| < 1;$$

$$\left|\langle (1_H + N)^{-1} x, y\rangle\right|$$
$$\leq \langle \left(1_H - |N|^{2\alpha}\right)^{-1} x, x\rangle^{1/2} \langle \left(1_H - |N|^{2\alpha}\right)^{-1} y, y\rangle^{1/2}, \|N\| < 1;$$

$$\left|\langle \sin (N) x, y\rangle\right|$$
$$\leq \langle \sinh \left(|N|^{2\alpha}\right) x, x\rangle^{1/2} \langle \sinh \left(|N|^{2(1-\alpha)}\right) y, y\rangle^{1/2}, \text{ for any } N;$$

$$\left|\langle \cos (N) x, y\rangle\right|$$
$$\leq \langle \cosh \left(|N|^{2\alpha}\right) x, x\rangle^{1/2} \langle \cosh \left(|N|^{2(1-\alpha)}\right) y, y\rangle^{1/2}, \text{ for any } N;$$

for any $x, y \in H$.

We have, for instance, the following norm inequalities as well:

$$\|\sin (N)\| \leq \left\|\sinh \left(|N|^{2\alpha}\right)\right\|^{1/2} \left\|\sinh \left(|N|^{2(1-\alpha)}\right)\right\|^{1/2};$$
$$\|\cos (N)\| \leq \left\|\cosh \left(|N|^{2\alpha}\right)\right\|^{1/2} \left\|\cosh \left(|N|^{2(1-\alpha)}\right)\right\|^{1/2}$$

for any normal operator N and

$$\left\|\ln \left(1_H + N\right)^{-1}\right\| \leq \left\|\ln \left(1_H - |N|^{2\alpha}\right)^{-1}\right\|^{1/2} \left\|\ln \left(1_H - |N|^{2\alpha}\right)^{-1}\right\|^{1/2}$$

for N with $\|N\| < 1$.

If we utilize the following function as power series representations with nonnegative coefficients:

$$\frac{1}{2} \ln \left(\frac{1+z}{1-z}\right) = \sum_{n=1}^{\infty} \frac{1}{2n-1} z^{2n-1}, \qquad z \in D(0, 1); \qquad (2.24)$$

$$\sin^{-1} (z) = \sum_{n=0}^{\infty} \frac{\Gamma\left(n + \frac{1}{2}\right)}{\sqrt{\pi} (2n+1) n!} z^{2n+1}, \qquad z \in D(0, 1);$$

$$\tanh^{-1} (z) = \sum_{n=1}^{\infty} \frac{1}{2n-1} z^{2n-1}, \qquad z \in D(0, 1)$$

$$_2F_1 (\alpha, \beta, \gamma, z) = \sum_{n=0}^{\infty} \frac{\Gamma(n+\alpha) \Gamma(n+\beta) \Gamma(\gamma)}{n! \Gamma(\alpha) \Gamma(\beta) \Gamma(n+\gamma)} z^n, \alpha, \beta, \gamma > 0,$$

$$z \in D(0, 1);$$

where Γ is the *Gamma function*, then we can state the following vector inequalities:

$$|\langle \exp(N) x, y \rangle| \tag{2.25}$$
$$\leq \langle \exp(|N|^{2\alpha}) x, x \rangle^{1/2} \langle \exp(|N|^{2(1-\alpha)}) y, y \rangle^{1/2} ;$$

$$\left| \left\langle \ln\left(\frac{1_H + N}{1_H - N}\right) x, y \right\rangle \right|$$
$$\leq \left\langle \ln\left(\frac{1_H + |N|^{2\alpha}}{1_H - |N|^{2\alpha}}\right) x, x \right\rangle^{1/2} \left\langle \ln\left(\frac{1_H + |N|^{2(1-\alpha)}}{1_H - |N|^{2(1-\alpha)}}\right) y, y \right\rangle^{1/2} ;$$

$$|\langle \sin^{-1}(N) x, y \rangle|$$
$$\leq \langle \sin^{-1}(|N|^{2\alpha}) x, x \rangle^{1/2} \langle \sin^{-1}(|N|^{2(1-\alpha)}) y, y \rangle^{1/2} ;$$

$$|\langle \tanh^{-1}(N) x, y \rangle|$$
$$\leq \langle \tanh^{-1}(|N|^{2\alpha}) x, x \rangle^{1/2} \langle \tanh^{-1}(|N|^{2(1-\alpha)}) y, y \rangle^{1/2} ;$$

$$|\langle {}_2F_1(\alpha, \beta, \gamma, N) x, y \rangle|$$
$$\leq \langle {}_2F_1(\alpha, \beta, \gamma, |N|^{2\alpha}) x, x \rangle^{1/2} \langle {}_2F_1(\alpha, \beta, \gamma, |N|^{2(1-\alpha)}) y, y \rangle^{1/2} ;$$

for any $x, y \in H$. The first inequality in (2.25) holds for any normal operator N while the other ones request the assumption $\|N\| < 1$.

We also have the norm inequalities

$$\|\exp(N)\| \leq \left\|\exp(|N|^{2\alpha})\right\|^{1/2} \left\|\exp(|N|^{2(1-\alpha)})\right\|^{1/2} ;$$
$$\|\cosh(N)\| \leq \left\|\cosh(|N|^{2\alpha})\right\|^{1/2} \left\|\cosh(|N|^{2(1-\alpha)})\right\|^{1/2} ;$$
$$\|\sinh(N)\| \leq \left\|\sinh(|N|^{2\alpha})\right\|^{1/2} \left\|\sinh(|N|^{2(1-\alpha)})\right\|^{1/2} ;$$

for any normal operator N and

$$\left\|\ln\left(\frac{1_H + N}{1_H - N}\right)\right\| \leq \left\|\ln\left(\frac{1_H + |N|^{2\alpha}}{1_H - |N|^{2\alpha}}\right)\right\|^{1/2} \left\|\ln\left(\frac{1_H + |N|^{2(1-\alpha)}}{1_H - |N|^{2(1-\alpha)}}\right)\right\|^{1/2}$$

for N with $\|N\| < 1$.

A similar result is the following one:

Theorem 2.8 (Dragomir [3]) *Let $f(z) = \sum_{n=0}^{\infty} a_n z^n$ be a function defined by power series with real coefficients and convergent on the open disk $D(0, R) \subset \mathbb{C}, R > 0$. If N is a normal operator on the Hilbert space H, $z \in \mathbb{C}$ such that $|z|^2$, $|z| \|N\|$,*

$\|N\|^2 < R$, then we have the inequalities

$$|\langle f (zN) x, y\rangle|^2 \leq f_A \left(|z|^2\right) \langle f_A \left(|N|^2\right) x, x\rangle^\alpha \langle f_A \left(|N|^2\right) y, y\rangle^{1-\alpha} \qquad (2.26)$$

for any $x, y \in H$ and $\alpha \in [0, 1]$.

In particular, we have

$$|\langle f (zN) x, y\rangle|^2 \leq f_A \left(|z|^2\right) \langle f_A \left(|N|^2\right) x, x\rangle^{1/2} \langle f_A \left(|N|^2\right) y, y\rangle^{1/2}. \qquad (2.27)$$

Proof By the Cauchy–Bunyakowsky–Schwarz inequality we have

$$\left|\left\langle \sum_{j=0}^n a_j z^j N^j x, y\right\rangle\right|^2 \leq \sum_{j=0}^n |a_j| |z|^{2j} \sum_{j=0}^n |a_j| |\langle N^j x, y\rangle|^2 \qquad (2.28)$$

for any $n \in \mathbb{N}$ and $x, y \in H$.

Utilising (2.5) we also have

$$\sum_{j=0}^n |a_j| |\langle N^j x, y\rangle|^2 \leq \left\langle \sum_{j=0}^n |a_j| |N^j|^2 x, x\right\rangle^\alpha \left\langle \sum_{j=0}^n |a_j| |N^j|^2 y, y\right\rangle^{1-\alpha} \qquad (2.29)$$

$$= \left\langle \sum_{j=0}^n |a_j| |N|^{2j} x, x\right\rangle^\alpha \left\langle \sum_{j=0}^n |a_j| |N|^{2j} y, y\right\rangle^{1-\alpha}$$

for any $n \in \mathbb{N}$ and $x, y \in H$.

By making use of (2.28) and (2.29) we get

$$\left|\left\langle \sum_{j=0}^n a_j z^j N^j x, y\right\rangle\right|^2 \qquad (2.30)$$

$$\leq \sum_{j=0}^n |a_j| |z|^{2j} \left\langle \sum_{j=0}^n |a_j| |N|^{2j} x, x\right\rangle^\alpha \left\langle \sum_{j=0}^n |a_j| |N|^{2j} y, y\right\rangle^{1-\alpha}$$

for any $n \in \mathbb{N}$ and $x, y \in H$.

Since the series $\sum_{j=0}^\infty |a_j| |N|^{2j}$ is absolutely convergent, taking the limit over $n \to \infty$ in (2.30) produces the desired result (2.26). ∎

Remark 2.9 Assume that f, R, z, N and α are as in Theorem 2.8. If we take the supremum in (2.26) over $y \in H$, $\|y\| = 1$, then we get

$$\|f (zN) x\|^2 \leq f_A \left(|z|^2\right) \langle f_A \left(|N|^2\right) x, x\rangle^\alpha \|f_A \left(|N|^2\right)\|^{1-\alpha} \qquad (2.31)$$

for any $x \in H$, which produces the operator norm inequality

$$\|f(zN)\|^2 \le f_A\left(|z|^2\right)\|f_A\left(|N|^2\right)\|. \qquad (2.32)$$

If we take $y = x$ in (2.26), then we get

$$|\langle f(zN)x, x\rangle|^2 \le f_A\left(|z|^2\right)\langle f_A\left(|N|^2\right)x, x\rangle \qquad (2.33)$$

for any $x \in H$.

From (2.26) we get the vector inequalities

$$|\langle \exp(zN)x, y\rangle|^2 \le \exp\left(|z|^2\right)\langle \exp\left(|N|^2\right)x, x\rangle^\alpha \langle \exp\left(|N|^2\right)y, y\rangle^{1-\alpha},$$

$$|\langle \sin(zN)x, y\rangle|^2 \le \sinh\left(|z|^2\right)\langle \sinh\left(|N|^2\right)x, x\rangle^\alpha \langle \sinh\left(|N|^2\right)y, y\rangle^{1-\alpha},$$

and

$$|\langle \cos(zN)x, y\rangle|^2$$
$$\le \cosh\left(|z|^2\right)\langle \cosh\left(|N|^2\right)x, x\rangle^\alpha \langle \cosh\left(|N|^2\right)y, y\rangle^{1-\alpha},$$

for any normal operator N, any complex number z and any $x, y \in H$.

We have, for instance, from (2.32) the following norm inequalities as well:

$$\|\exp(zN)\|^2 \le \exp\left(|z|^2\right)\|\exp\left(|N|^2\right)\|$$

and

$$\|\sin(zN)\|^2 \le \sinh\left(|z|^2\right)\|\sinh\left(|N|^2\right)\|$$

for any normal operator N and any complex number z.

Similar results can be stated for other functions, however the details are omitted.

2.3 Inequalities for the Euclidian Norm

In [30], the author has introduced the following norm on the Cartesian product $B^{(n)}(H) := B(H) \times \cdots \times B(H)$, where $B(H)$ denotes the Banach algebra of all bounded linear operators defined on the complex Hilbert space H:

$$\|(T_1, \ldots, T_n)\|_e := \sup_{(\lambda_1, \ldots, \lambda_n) \in \mathbb{B}_n} \|\lambda_1 T_1 + \cdots + \lambda_n T_n\|, \qquad (2.34)$$

where $(T_1, \ldots, T_n) \in \mathcal{B}^{(n)}(H)$ and $\mathbb{B}_n := \left\{ (\lambda_1, \ldots, \lambda_n) \in \mathbb{C}^n \left| \sum_{j=1}^n |\lambda_j|^2 \le 1 \right. \right\}$ is the Euclidean closed ball in $\mathbb{C}^n$.

It is clear that $\|\cdot\|_e$ is a norm on $\mathcal{B}^{(n)}(H)$ and for any $(T_1, \ldots, T_n) \in \mathcal{B}^{(n)}(H)$ we have

$$\|(T_1, \ldots, T_n)\|_e = \|(T_1^*, \ldots, T_n^*)\|_e,$$

where T_j^* is the adjoint operator of T_j, $j \in \{1, \ldots, n\}$. We call this the *Euclidian norm* of an n-tuple of operators $(T_1, \ldots, T_n) \in \mathcal{B}^{(n)}(H)$.

It has been shown in [30] that the following basic inequality for the Euclidian norm holds true:

$$\frac{1}{\sqrt{n}} \left\| \sum_{j=1}^n |T_j^*|^2 \right\|^{\frac{1}{2}} \le \|(T_1, \ldots, T_n)\|_e \le \left\| \sum_{j=1}^n |T_j^*|^2 \right\|^{\frac{1}{2}} \tag{2.35}$$

for any n-tuple $(T_1, \ldots, T_n) \in \mathcal{B}^{(n)}(H)$ and the constants $\frac{1}{\sqrt{n}}$ and 1 are best possible.

In the same paper [30] the author has introduced the *Euclidean operator radius* of an n-tuple of operators $(T_1, \ldots, T_n)$ by

$$w_e(T_1, \ldots, T_n) := \sup_{\|x\|=1} \left(\sum_{j=1}^n |\langle T_j x, x \rangle|^2 \right)^{\frac{1}{2}} \tag{2.36}$$

and proved that $w_e(\cdot)$ is a norm on $\mathcal{B}^{(n)}(H)$ and satisfies the double inequality:

$$\frac{1}{2} \|(T_1, \ldots, T_n)\|_e \le w_e(T_1, \ldots, T_n) \le \|(T_1, \ldots, T_n)\|_e \tag{2.37}$$

for each n-tuple $(T_1, \ldots, T_n) \in \mathcal{B}^{(n)}(H)$.

As pointed out in [30], the Euclidean numerical radius also satisfies the double inequality:

$$\frac{1}{2\sqrt{n}} \left\| \sum_{j=1}^n |T_j^*|^2 \right\|^{\frac{1}{2}} \le w_e(T_1, \ldots, T_n) \le \left\| \sum_{j=1}^n |T_j^*|^2 \right\|^{\frac{1}{2}} \tag{2.38}$$

for any $(T_1, \ldots, T_n) \in \mathcal{B}^{(n)}(H)$ and the constants $\frac{1}{2\sqrt{n}}$ and 1 are best possible.

In [2], by utilizing the concept of *hypo-Euclidean norm* on H^n we obtained the following representation for the Euclidian norm:

Proposition 2.10 (Dragomir [2]) *For any* $(T_1, \ldots, T_n) \in \mathcal{B}^{(n)}(H)$ *we have*

$$\|(T_1, \ldots, T_n)\|_e = \sup_{\|y\|=1, \|x\|=1} \left(\sum_{j=1}^{n} |\langle T_j y, x \rangle|^2 \right)^{\frac{1}{2}}. \tag{2.39}$$

The following different lower bound for the Euclidean operator norm $\|\cdot\|_e$ was also obtained in [2]:

Proposition 2.11 (Dragomir [2]) *For any* $(T_1, \ldots, T_n) \in B^{(n)}(H)$, *we have*

$$\|(T_1, \ldots, T_n)\|_e \geq \frac{1}{\sqrt{n}} \|T_1 + \cdots + T_n\|. \tag{2.40}$$

Utilizing some techniques based on the Boas-Bellman and Bombieri type inequalities we obtained in [2] the following upper bounds:

Proposition 2.12 (Dragomir [2]) *For any* $(T_1, \ldots, T_n) \in B^{(n)}(H)$, *we have the inequalities*:

$$\|(T_1, \ldots, T_n)\|_e^2 \leq \begin{cases} \max_{1 \leq j \leq n} \left\{ \|T_j\|^2 \right\} + \left[\sum_{1 \leq j \neq k \leq n} w^2 \left(T_k^* T_j \right) \right]^{\frac{1}{2}}; \\ \max_{1 \leq j \leq n} \left\{ \|T_j\|^2 \right\} + (n-1) \max_{1 \leq j \neq k \leq n} \left\{ w \left(T_k^* T_j \right) \right\}; \\ \left[\max_{1 \leq j \leq n} \left\{ \|T_j\|^2 \right\} \left\| \sum_{j=1}^{n} |T_j|^2 \right\|^2 \\ + \max_{1 \leq j \neq k \leq n} \left\{ \|T_j\| \|T_k\| \right\} \sum_{1 \leq j \neq k \leq n} w \left(T_k T_j^* \right) \right]^{\frac{1}{2}} \end{cases} \tag{2.41}$$

and

$$\|(T_1, \ldots, T_n)\|_e^2 \leq \begin{cases} \max_{1 \leq j \leq n} \left\{ \sum_{k=1}^{n} w \left(T_k^* T_j \right) \right\}; \\ \left[\sum_{j,k=1}^{n} w^2 \left(T_k^* T_j \right) \right]^{\frac{1}{2}}; \\ n \max_{1 \leq j \leq n} \left[\sum_{k=1}^{n} w^2 \left(T_k^* T_j \right) \right]^{\frac{1}{2}}; \\ n \left[\sum_{j=1}^{n} \max_{1 \leq k \leq n} \left\{ w^2 \left(T_k^* T_j \right) \right\} \right]^{\frac{1}{2}}. \end{cases} \tag{2.42}$$

Now we can provide now a different upper bound for the Euclidian norm:

Proposition 2.13 (Dragomir [3]) *Let* $(T_1, \ldots, T_n) \in B^{(n)}(H)$ *be an n-tuple of bounded linear operators on the Hilbert space* $(H; \langle \cdot, \cdot \rangle)$. *Then we have*

$$\|(T_1, \dots, T_n)\|_e^2 \le \left\|\sum_{j=1}^{n} |T_j|^2\right\|^{\alpha} \left\|\sum_{j=1}^{n} |T_j^*|^2\right\|^{1-\alpha} \tag{2.43}$$

and

$$w_e^2(T_1, \dots, T_n) \le \sup_{\|x\|=1} \left[\left\langle \sum_{j=1}^{n} |T_j|^2 x, x\right\rangle^{\alpha} \left\langle \sum_{j=1}^{n} |T_j^*|^2 x, x\right\rangle^{1-\alpha}\right] \tag{2.44}$$

$$\le \begin{cases} \left[\left\|\sum_{j=1}^{n} |T_j|^2\right\|\right]^{\alpha} \left[\left\|\sum_{j=1}^{n} |T_j^*|^2\right\|\right]^{1-\alpha}, \\[2ex] \left\|\sum_{j=1}^{n} \left[\alpha |T_j|^2 + (1-\alpha) |T_j^*|^2\right]\right\|, \end{cases}$$

for any $\alpha \in [0, 1]$.

Proof Utilizing the vector inequality (2.1) and taking the supremum over $\|y\| = 1$, $\|x\| = 1$ we have

$$\|(T_1, \dots, T_n)\|_e^2 \le \left[\sup_{\|x\|=1} \left\langle \sum_{j=1}^{n} |T_j|^2 x, x\right\rangle\right]^{\alpha} \left[\sup_{\|y\|=1} \left\langle \sum_{j=1}^{n} |T_j^*|^2 y, y\right\rangle\right]^{1-\alpha} \tag{2.45}$$

for any $\alpha \in [0, 1]$ and since

$$\sup_{\|x\|=1} \left\langle \sum_{j=1}^{n} |T_j|^2 x, x\right\rangle = \left\|\sum_{j=1}^{n} |T_j|^2\right\|$$

and

$$\sup_{\|y\|=1} \left\langle \sum_{j=1}^{n} |T_j^*|^2 y, y\right\rangle = \left\|\sum_{j=1}^{n} |T_j^*|^2\right\|$$

we get from (2.45) the desired result (2.43).

Now from the first inequality in (2.4) we have

$$w_e^2(T_1, \dots, T_n) \le \sup_{\|x\|=1} \left[\left\langle \sum_{j=1}^{n} |T_j|^2 x, x\right\rangle^{\alpha} \left\langle \sum_{j=1}^{n} |T_j^*|^2 x, x\right\rangle^{1-\alpha}\right] \tag{2.46}$$

$$\le \left[\sup_{\|x\|=1} \left\langle \sum_{j=1}^{n} |T_j|^2 x, x\right\rangle\right]^{\alpha} \left[\sup_{\|x\|=1} \left\langle \sum_{j=1}^{n} |T_j^*|^2 x, x\right\rangle\right]^{1-\alpha}$$

$$= \left[\left\| \sum_{j=1}^{n} |T_j|^2 \right\| \right]^{\alpha} \left[\left\| \sum_{j=1}^{n} |T_j^*|^2 \right\| \right]^{1-\alpha}$$

and from the second inequality in (2.4) we also have

$$w_e^2 (T_1, \ldots, T_n) \le \sup_{\|x\|=1} \left\langle \sum_{j=1}^{n} \left[\alpha |T_j|^2 + (1-\alpha) |T_j^*|^2 \right] x, x \right\rangle \qquad (2.47)$$

$$= \left\| \sum_{j=1}^{n} \alpha |T_j|^2 + (1-\alpha) |T_j^*|^2 \right\|$$

for any $\alpha \in [0, 1]$.

Utilizing (2.46) and (2.47) we get (2.44).   ∎

Remark 2.14 The case when $\alpha = 1/2$ provides the inequalities

$$\| (T_1, \ldots, T_n) \|_e^2 \le \left\| \sum_{j=1}^{n} |T_j|^2 \right\|^{1/2} \left\| \sum_{j=1}^{n} |T_j^*|^2 \right\|^{1/2} \qquad (2.48)$$

and

$$w_e^2 (T_1, \ldots, T_n) \le \sup_{\|x\|=1} \left[\left\langle \sum_{j=1}^{n} |T_j|^2 x, x \right\rangle^{1/2} \left\langle \sum_{j=1}^{n} |T_j^*|^2 x, x \right\rangle^{1/2} \right] \qquad (2.49)$$

$$\le \begin{cases} \left[\left\| \sum_{j=1}^{n} |T_j|^2 \right\| \right]^{1/2} \left[\left\| \sum_{j=1}^{n} |T_j^*|^2 \right\| \right]^{1/2}, \\ \left\| \sum_{j=1}^{n} \left[\frac{|T_j|^2 + |T_j^*|^2}{2} \right] \right\|. \end{cases}$$

2.4 Inequalities for s-1-Norm and s-1-Numerical Radius

We can introduce the *s-p-norm* of the *n*-tuple of operators $(T_1, \ldots, T_n) \in B^{(n)}(H)$ by

$$\| (T_1, \ldots, T_n) \|_{s,p} := \sup_{\|y\|=1, \|x\|=1} \left[\left(\sum_{j=1}^{n} |\langle T_j y, x \rangle|^p \right)^{\frac{1}{p}} \right]. \qquad (2.50)$$

Indeed this is a norm, since by the Minkowski inequality we have

$$\|(T_1, \ldots, T_n) + (V_1, \ldots, V_n)\|_{s,p} \qquad (2.51)$$

$$= \sup_{\|y\|=1, \|x\|=1} \left[\left(\sum_{j=1}^{n} |\langle T_j y, x \rangle + \langle V_j y, x \rangle|^p \right)^{\frac{1}{p}} \right]$$

$$\leq \sup_{\|y\|=1, \|x\|=1} \left[\left(\sum_{j=1}^{n} |\langle T_j y, x \rangle|^p \right)^{\frac{1}{p}} + \left(\sum_{j=1}^{n} |\langle V_j y, x \rangle|^p \right)^{\frac{1}{p}} \right]$$

$$\leq \sup_{\|y\|=1, \|x\|=1} \left(\sum_{j=1}^{n} |\langle T_j y, x \rangle|^p \right)^{\frac{1}{p}} + \sup_{\|y\|=1, \|x\|=1} \left(\sum_{j=1}^{n} |\langle V_j y, x \rangle|^p \right)^{\frac{1}{p}}$$

$$= \|(T_1, \ldots, T_n)\|_{s,p} + \|(V_1, \ldots, V_n)\|_{s,p} ,$$

which proves the triangle inequality. The other properties of the norm are obvious.
For $p = 2$ we get

$$\|(T_1, \ldots, T_n)\|_{s,2} = \|(T_1, \ldots, T_n)\|_e .$$

We are interested in this section in the case $p = 1$, namely on the s-1-norm defined by

$$\|(T_1, \ldots, T_n)\|_{s,1} := \sup_{\|y\|=1, \|x\|=1} \sum_{j=1}^{n} |\langle T_j y, x \rangle| .$$

Since for any $x, y \in H$ we have $\sum_{j=1}^{n} |\langle T_j y, x \rangle| \geq \left| \langle \sum_{j=1}^{n} T_j y, x \rangle \right|$, then by the properties of the supremum we get the basic inequality

$$\left\| \sum_{j=1}^{n} T_j \right\| \leq \|(T_1, \ldots, T_n)\|_{s,1} \leq \sum_{j=1}^{n} \|T_j\| . \qquad (2.52)$$

Similarly, we can also introduce the s-p-numerical radius of the n-tuple of operators $(T_1, \ldots, T_n) \in B^{(n)}(H)$ by

$$w_{s,p}(T_1, \ldots, T_n) := \sup_{\|x\|=1} \left[\left(\sum_{j=1}^{n} |\langle T_j x, x \rangle|^p \right)^{\frac{1}{p}} \right] , \qquad (2.53)$$

which for $p = 2$ reduces to the Euclidean operator radius introduced previously. We observe that the s-p-numerical radius is also a norm on $B^{(n)}(H)$ for $p \geq 1$ and for $p = 1$ it satisfies the basic inequality

$$w\left(\sum_{j=1}^{n} T_j\right) \leq w_{s,1}(T_1, \ldots, T_n) \leq \sum_{j=1}^{n} w(T_j). \tag{2.54}$$

Proposition 2.15 (Dragomir [3]) *Let* $(T_1, \ldots, T_n) \in \mathcal{B}^{(n)}(H)$ *be an n-tuple of bounded linear operators on the Hilbert space* $(H; \langle \cdot, \cdot \rangle)$. *Then we have*

$$\|(T_1, \ldots, T_n)\|_{s,1} \leq \left\|\sum_{j=1}^{n} |T_j|^{2\alpha}\right\|^{1/2} \left\|\sum_{j=1}^{n} |T_j^*|^{2(1-\alpha)}\right\|^{1/2} \tag{2.55}$$

for any $\alpha \in [0, 1]$, *and in particular, the following refinement of the triangle inequality for operator norm:*

$$\left\|\sum_{j=1}^{n} T_j\right\| \leq \|(T_1, \ldots, T_n)\|_{s,1} \tag{2.56}$$

$$\leq \left\|\sum_{j=1}^{n} |T_j|\right\|^{1/2} \left\|\sum_{j=1}^{n} |T_j^*|\right\|^{1/2}$$

$$\leq \frac{1}{2}\left[\left\|\sum_{j=1}^{n} |T_j|\right\| + \left\|\sum_{j=1}^{n} |T_j^*|\right\|\right] \leq \sum_{j=1}^{n} \|T_j\|.$$

Proof Utilizing the vector inequality (2.7) and taking the supremum over $\|y\| = 1$, $\|x\| = 1$ we have

$$\|(T_1, \ldots, T_n)\|_{s,1} \tag{2.57}$$

$$\leq \left\{\sup_{\|x\|=1} \left\langle \sum_{j=1}^{n} |T_j|^{2\alpha} x, x\right\rangle\right\}^{1/2} \left\{\sup_{\|y\|=1} \left\langle \sum_{j=1}^{n} |T_j^*|^{2(1-\alpha)} y, y\right\rangle\right\}^{1/2}$$

and since

$$\sup_{\|x\|=1} \left\langle \sum_{j=1}^{n} |T_j|^{2\alpha} x, x\right\rangle = \left\|\sum_{j=1}^{n} |T_j|^{2\alpha}\right\|$$

and

$$\sup_{\|y\|=1} \left\langle \sum_{j=1}^{n} |T_j^*|^{2(1-\alpha)} y, y\right\rangle = \left\|\sum_{j=1}^{n} |T_j^*|^{2(1-\alpha)}\right\|$$

then we get from (2.57) the desired inequality (2.55).

The inequality (2.56) follows from (2.55).   ∎

The case of normal operators provides a simpler bound:

Corollary 2.16 (Dragomir [3]) *Let* $(N_1, ..., N_n) \in \mathcal{B}^{(n)}(H)$ *be an n-tuple of normal operators on the Hilbert space* $(H; \langle \cdot, \cdot \rangle)$. *Then we have*

$$\|(N_1, \ldots, N_n)\|_{s,1} \leq \left\| \sum_{j=1}^{n} |N_j|^{2\alpha} \right\|^{1/2} \left\| \sum_{j=1}^{n} |N_j|^{2(1-\alpha)} \right\|^{1/2} \qquad (2.58)$$

for any $\alpha \in [0, 1]$, *and in particular,*

$$\left\| \sum_{j=1}^{n} N_j \right\| \leq \|(N_1, \ldots, N_n)\|_{s,1} \leq \left\| \sum_{j=1}^{n} |N_j| \right\| \leq \sum_{j=1}^{n} \|N_j\|. \qquad (2.59)$$

The above results provide an interesting criterion of convergence in the Banach algebra $\mathcal{B}(H)$ for the series of operators $\sum_{j=0}^{\infty} T_j$.

Criterion 2.17 (Dragomir [3]) *Let* $\{T_j\}_{j \in \mathbb{N}}$ *be a sequence of operators in* $\mathcal{B}(H)$. *If there exists an* $\alpha \in (0, 1)$ *such that the series* $\sum_{j=0}^{\infty} |T_j|^{2\alpha}$ *and* $\sum_{j=0}^{\infty} |T_j^*|^{2(1-\alpha)}$ *are convergent in the Banach algebra* $\mathcal{B}(H)$, *then* $\sum_{j=0}^{\infty} T_j$ *is convergent in* $\mathcal{B}(H)$ *and*

$$\left\| \sum_{j=0}^{\infty} T_j \right\| \leq \left\| \sum_{j=0}^{\infty} |T_j|^{2\alpha} \right\|^{1/2} \left\| \sum_{j=0}^{\infty} |T_j^*|^{2(1-\alpha)} \right\|^{1/2}.$$

In particular, the convergence of the series $\sum_{j=0}^{\infty} |T_j|$ *and* $\sum_{j=0}^{\infty} |T_j^*|$ *imply the convergence of* $\sum_{j=0}^{\infty} T_j$ *in* $\mathcal{B}(H)$ *with the estimate for the sums as follows:*

$$\left\| \sum_{j=0}^{\infty} T_j \right\| \leq \left\| \sum_{j=0}^{\infty} |T_j| \right\|^{1/2} \left\| \sum_{j=0}^{\infty} |T_j^*| \right\|^{1/2}.$$

The following result for the s-1-numerical radius may be stated as well:

Proposition 2.18 (Dragomir [3]) *Let* $(T_1, ..., T_n) \in \mathcal{B}^{(n)}(H)$ *be an n-tuple of bounded linear operators on the Hilbert space* $(H; \langle \cdot, \cdot \rangle)$. *Then we have*

$$w_{s,1}(T_1, \ldots, T_n) \qquad (2.60)$$

$$\leq \sup \left\{ \left\langle \sum_{j=1}^{n} p_j \left| T_j \right|^{2\alpha} x, x \right\rangle^{1/2} \left\langle \sum_{j=1}^{n} p_j \left| T_j^* \right|^{2(1-\alpha)} x, x \right\rangle^{1/2} \right\}$$

$$\leq \begin{cases} \left\| \sum_{j=1}^{n} \left| T_j \right|^{2\alpha} \right\|^{1/2} \left\| \sum_{j=1}^{n} \left| T_j^* \right|^{2(1-\alpha)} \right\|^{1/2} ; \\[2mm] \left\| \sum_{j=1}^{n} \dfrac{\left| T_j \right|^{2\alpha} + \left| T_j^* \right|^{2(1-\alpha)}}{2} \right\| \end{cases}$$

for any $\alpha \in [0, 1]$, and, in particular,

$$w \left(\sum_{j=1}^{n} T_j \right) \leq w_{s,1}(T_1, \ldots, T_n) \qquad (2.61)$$

$$\leq \begin{cases} \left\| \sum_{j=1}^{n} \left| T_j \right| \right\|^{1/2} \left\| \sum_{j=1}^{n} \left| T_j^* \right| \right\|^{1/2} ; \\[2mm] \left\| \sum_{j=1}^{n} \dfrac{\left| T_j \right| + \left| T_j^* \right|}{2} \right\|. \end{cases}$$

Remark 2.19 We observe that due to the inequality

$$\frac{1}{2} \left\| \sum_{j=1}^{n} T_j \right\| \leq w \left(\sum_{j=1}^{n} T_j \right) \leq \left\| \sum_{j=1}^{n} \frac{\left| T_j \right|^{2\alpha} + \left| T_j^* \right|^{2(1-\alpha)}}{2} \right\|, \qquad (2.62)$$

the convergence of the series $\sum_{k=0}^{\infty} \left[\left| T_k \right|^{2\alpha} + \left| T_k^* \right|^{2(1-\alpha)} \right]$ in the Banach algebra $\mathcal{B}(H)$

for some $\alpha \in (0, 1)$ suffices for the convergence of $\sum_{k=0}^{\infty} T_k$, which is a slight improve-

ment of the result from Criterion 2.17.

The case $\alpha = \frac{1}{2}$ produces the simpler inequality of interest for the numerical radius of a sum:

$$\frac{1}{2} \left\| \sum_{j=1}^{n} T_j \right\| \leq w \left(\sum_{j=1}^{n} T_j \right) \leq \frac{1}{2} \left\| \sum_{j=1}^{n} \left[\left| T_j \right| + \left| T_j^* \right| \right] \right\|. \qquad (2.63)$$

2.5 Additive Inequalities

Employing the original Kato's inequality we can state the following new result:

Theorem 2.20 (Dragomir et al. [9]) *Let* $(T_1, ..., T_n) \in \mathcal{B}^{(n)}(H)$ *be an n-tuple of bounded linear operators on the Hilbert space* $(H; \langle \cdot, \cdot \rangle)$ *and* $(p_1, ..., p_n) \in \mathbb{R}_+^{*n}$ *an n-tuple of nonnegative weights not all of them equal to zero. Then we have*

$$\sum_{j=1}^{n} p_j \left| \langle T_j x, y \rangle \right| \leq \left\langle \sum_{j=1}^{n} p_j \left(\frac{|T_j|^{2\alpha} + |T_j|^{2(1-\alpha)}}{2} \right) x, x \right\rangle^{1/2} \tag{2.64}$$

$$\times \left\langle \sum_{j=1}^{n} p_j \left(\frac{\left|T_j^*\right|^{2\alpha} + \left|T_j^*\right|^{2(1-\alpha)}}{2} \right) y, y \right\rangle^{1/2}$$

for any $x, y \in H$, $\alpha \in [0, 1]$ *and, in particular, for* $\alpha = \frac{1}{2}$

$$\sum_{j=1}^{n} p_j \left| \langle T_j x, y \rangle \right| \leq \left\langle \sum_{j=1}^{n} p_j |T_j| x, x \right\rangle^{1/2} \left\langle \sum_{j=1}^{n} p_j |T_j^*| y, y \right\rangle^{1/2} \tag{2.65}$$

for any $x, y \in H$.

Proof Utilising Kato's inequality we have

$$\left| \langle T_j x, y \rangle \right| \leq \left\langle |T_j|^{2\alpha} x, x \right\rangle^{1/2} \left\langle |T_j^*|^{2(1-\alpha)} y, y \right\rangle^{1/2}$$

and, by replacing α with $1 - \alpha$,

$$\left| \langle T_j x, y \rangle \right| \leq \left\langle |T_j|^{2(1-\alpha)} x, x \right\rangle^{1/2} \left\langle |T_j^*|^{2\alpha} y, y \right\rangle^{1/2},$$

which, by summation gives

$$\left| \langle T_j x, y \rangle \right| \leq \frac{1}{2} \left[\left\langle |T_j|^{2\alpha} x, x \right\rangle^{1/2} \left\langle |T_j^*|^{2(1-\alpha)} y, y \right\rangle^{1/2} \tag{2.66} \right.$$

$$\left. + \left\langle |T_j|^{2(1-\alpha)} x, x \right\rangle^{1/2} \left\langle |T_j^*|^{2\alpha} y, y \right\rangle^{1/2} \right]$$

for any $j \in \{1, ..., n\}$ and $x, y \in H$.

By the elementary inequality

$$ab + cd \leq \left(a^2 + c^2\right)^{1/2} \left(b^2 + d^2\right)^{1/2}, a, b, c, d \geq 0 \tag{2.67}$$

we have

$$\left[\left\langle |T_j|^{2\alpha} x, x\right\rangle^{1/2} \left\langle |T_j^*|^{2(1-\alpha)} y, y\right\rangle^{1/2} + \left\langle |T_j|^{2(1-\alpha)} x, x\right\rangle^{1/2} \left\langle |T_j^*|^{2\alpha} y, y\right\rangle^{1/2}\right]$$

$$\leq \left[\left\langle \left(|T_j|^{2\alpha} + |T_j|^{2(1-\alpha)}\right) x, x\right\rangle\right]^{1/2} \left[\left\langle \left(|T_j^*|^{2\alpha} + |T_j^*|^{2(1-\alpha)}\right) y, y\right\rangle\right]^{1/2},$$

which, by (2.66), produces

$$|\langle T_j x, y\rangle| \leq \left\langle \left(\frac{|T_j|^{2\alpha} + |T_j|^{2(1-\alpha)}}{2}\right) x, x\right\rangle^{1/2} \tag{2.68}$$

$$\times \left\langle \left(\frac{\left|T_j^*\right|^{2\alpha} + \left|T_j^*\right|^{2(1-\alpha)}}{2}\right) y, y\right\rangle^{1/2}$$

for any $j \in \{1, ..., n\}$ and $x, y \in H$.

Multiplying the inequalities (2.68) with the positive weights p_j, summing over j from 1 to n and utilizing the weighted Cauchy–Buniakowski–Schwarz inequality

$$\sum_{j=1}^{n} p_j a_j b_j \leq \left(\sum_{j=1}^{n} p_j a_j^2\right)^{1/2} \left(\sum_{j=1}^{n} p_j b_j^2\right)^{1/2}$$

where $(a_1, ..., a_n), (b_1, ..., b_n) \in \mathbb{R}_+^n$, we have

$$\sum_{j=1}^{n} p_j |\langle T_j x, y\rangle| \leq \sum_{j=1}^{n} p_j \left(\left\langle \left(\frac{|T_j|^{2\alpha} + |T_j|^{2(1-\alpha)}}{2}\right) x, x\right\rangle^{1/2}\right) \tag{2.69}$$

$$\times \left\langle \left(\frac{\left|T_j^*\right|^{2\alpha} + \left|T_j^*\right|^{2(1-\alpha)}}{2}\right) y, y\right\rangle^{1/2}$$

$$\leq \left\langle \sum_{j=1}^{n} p_j \left(\frac{|T_j|^{2\alpha} + |T_j|^{2(1-\alpha)}}{2}\right) x, x\right\rangle^{1/2}$$

$$\times \left\langle \sum_{j=1}^{n} p_j \left(\frac{\left|T_j^*\right|^{2\alpha} + \left|T_j^*\right|^{2(1-\alpha)}}{2}\right) y, y\right\rangle^{1/2}$$

for any $j \in \{1, ..., n\}$ and $x, y \in H$, and the inequality in (2.64) is proved. ∎

For vectors of norm one, the second inequality from (2.64) and (2.65) can be refined as follows:

Remark 2.21 With the assumptions in Theorem 2.20 we have

$$\sum_{j=1}^{n} p_j \left| \langle T_j x, y \rangle \right| \leq \left\langle \sum_{j=1}^{n} p_j \left(\frac{|T_j|^{2\alpha} + |T_j|^{2(1-\alpha)}}{2} \right) x, x \right\rangle^{1/2} \tag{2.70}$$

$$\times \left\langle \sum_{j=1}^{n} p_j \left(\frac{\left|T_j^*\right|^{2\alpha} + \left|T_j^*\right|^{2(1-\alpha)}}{2} \right) y, y \right\rangle^{1/2}$$

$$\leq \left\langle \left[\sum_{j=1}^{n} p_j \left(\frac{|T_j|^{2\alpha} + |T_j|^{2(1-\alpha)}}{2} \right) \right]^{1/2} x, x \right\rangle$$

$$\times \left\langle \left[\sum_{j=1}^{n} p_j \left(\frac{\left|T_j^*\right|^{2\alpha} + \left|T_j^*\right|^{2(1-\alpha)}}{2} \right) \right]^{1/2} y, y \right\rangle$$

$$\leq \frac{1}{2} \left[\left\langle \left[\sum_{j=1}^{n} p_j \left(\frac{|T_j|^{2\alpha} + |T_j|^{2(1-\alpha)}}{2} \right) \right]^{1/2} x, x \right\rangle^{2} \right.$$

$$\left. + \left\langle \left[\sum_{j=1}^{n} p_j \left(\frac{\left|T_j^*\right|^{2\alpha} + \left|T_j^*\right|^{2(1-\alpha)}}{2} \right) \right]^{1/2} y, y \right\rangle^{2} \right]$$

$$\leq \frac{1}{2} \left[\left\langle \sum_{j=1}^{n} p_j \left(\frac{|T_j|^{2\alpha} + |T_j|^{2(1-\alpha)}}{2} \right) x, x \right\rangle \right.$$

$$\left. + \left\langle \sum_{j=1}^{n} p_j \left(\frac{\left|T_j^*\right|^{2\alpha} + \left|T_j^*\right|^{2(1-\alpha)}}{2} \right) y, y \right\rangle \right]$$

for any $x, y \in H$ with $\|x\| = \|y\| = 1$.

In particular, we have

$$\sum_{j=1}^{n} p_j \left| \langle T_j x, y \rangle \right| \tag{2.71}$$

$$\leq \left\langle \sum_{j=1}^{n} p_j |T_j| x, x \right\rangle^{1/2} \left\langle \sum_{j=1}^{n} p_j |T_j^*| y, y \right\rangle^{1/2}$$

$$\leq \left\langle \left(\sum_{j=1}^{n} p_j \left|T_j\right|\right)^{1/2} x, x \right\rangle \left\langle \left(\sum_{j=1}^{n} p_j \left|T_j^*\right|\right)^{1/2} y, y \right\rangle$$

$$\leq \frac{1}{2} \left[\left\langle \left(\sum_{j=1}^{n} p_j \left|T_j\right|\right)^{1/2} x, x \right\rangle^2 + \left\langle \left(\sum_{j=1}^{n} p_j \left|T_j^*\right|\right)^{1/2} y, y \right\rangle^2 \right]$$

$$\leq \frac{1}{2} \left[\left\langle \sum_{j=1}^{n} p_j \left|T_j\right| x, x \right\rangle + \left\langle \sum_{j=1}^{n} p_j \left|T_j^*\right| y, y \right\rangle \right]$$

for any $x, y \in H$ with $\|x\| = \|y\| = 1$.

The proof follow by utilizing the Hölder-McCarthy inequalities (see for instance [29]) $\langle P^r x, x \rangle \leq \langle P x, x \rangle^r$ and $\langle P x, x \rangle^s \leq \langle P^s x, x \rangle$ that hold for the positive operator P, for $r \in (0, 1)$, $s \in [1, \infty)$ and $x \in H$ with $\|x\| = 1$. The details are omitted.

Remark 2.22 We observe also that the choice $y = x$ in the inequality (2.71) produces the result

$$\sum_{j=1}^{n} \left|\langle T_j x, x \rangle\right| \leq \left\langle \sum_{j=1}^{n} \left|T_j\right| x, x \right\rangle^{1/2} \left\langle \sum_{j=1}^{n} \left|T_j^*\right| x, x \right\rangle^{1/2} \tag{2.72}$$

$$\leq \left\langle \left(\sum_{j=1}^{n} \left|T_j\right|\right)^{1/2} x, x \right\rangle \left\langle \left(\sum_{j=1}^{n} \left|T_j^*\right|\right)^{1/2} x, x \right\rangle$$

$$\leq \frac{1}{2} \left[\left\langle \left(\sum_{j=1}^{n} \left|T_j\right|\right)^{1/2} x, x \right\rangle^2 + \left\langle \left(\sum_{j=1}^{n} \left|T_j^*\right|\right)^{1/2} x, x \right\rangle^2 \right]$$

$$\leq \left\langle \sum_{j=1}^{n} \left[\frac{\left|T_j\right| + \left|T_j^*\right|}{2} \right] x, x \right\rangle$$

for any $\in H$ with $\|x\| = 1$.

Remark 2.23 In order to provide some applications for functions of normal operators defined by power series, we need to state the inequality (2.64) for normal operators N_j, $j \in \{1, ..., n\}$, namely

$$\sum_{j=1}^{n} p_j \left|\langle N_j x, y\rangle\right| \le \left\langle \sum_{j=1}^{n} p_j \left(\frac{|N_j|^{2\alpha} + |N_j|^{2(1-\alpha)}}{2}\right) x, x \right\rangle^{1/2} \tag{2.73}$$

$$\times \left\langle \sum_{j=1}^{n} p_j \left(\frac{|N_j|^{2\alpha} + |N_j|^{2(1-\alpha)}}{2}\right) y, y \right\rangle^{1/2}$$

for any $\alpha \in [0, 1]$ and for any $x, y \in H$.

From a different perspective that involves quadratics, we can state the following result as well:

Theorem 2.24 (Dragomir et al. [9]) *Let* $(T_1, ..., T_n) \in \mathcal{B}^{(n)}(H)$ *be an n-tuple of bounded linear operators on the Hilbert space* $(H; \langle \cdot, \cdot \rangle)$ *and* $(p_1, ..., p_n) \in \mathbb{R}_+^{*n}$ *an n-tuple of nonnegative weights not all of them equal to zero. Then we have*

$$\sum_{j=1}^{n} p_j \left|\langle T_j x, y\rangle\right|^2 \tag{2.74}$$

$$\le \frac{1}{2} \sum_{j=1}^{n} p_j \left(\|T_j x\|^{2\alpha} \|T_j^* y\|^{2(1-\alpha)} + \|T_j^* y\|^{2\alpha} \|T_j x\|^{2(1-\alpha)}\right)$$

$$\le \frac{1}{2} \left[\left(\sum_{j=1}^{n} p_j \|T_j x\|^2\right)^{\alpha} \left(\sum_{j=1}^{n} p_j \|T_j^* y\|^2\right)^{1-\alpha} \right.$$

$$\left. + \left(\sum_{j=1}^{n} p_j \|T_j x\|^2\right)^{1-\alpha} \left(\sum_{j=1}^{n} p_j \|T_j^* y\|^2\right)^{\alpha} \right]$$

$$\le \frac{1}{2} \sum_{j=1}^{n} p_j \left(\|T_j x\|^2 + \|T_j^* y\|^2\right)$$

for any $x, y \in H$ *with* $\|x\| = \|y\| = 1$ *and* $\alpha \in [0, 1]$.

Proof We must prove the inequalities only in the case $\alpha \in (0, 1)$, since the case $\alpha = 0$ or $\alpha = 1$ follows directly from the corresponding case of Kato's inequality.

Utilizing Kato's inequality for the operator T_j, $j \in \{1, ..., n\}$ we have

$$\left|\langle T_j x, y\rangle\right|^2 \le \left\langle |T_j|^{2\alpha} x, x\right\rangle \left\langle |T_j^*|^{2(1-\alpha)} y, y\right\rangle \tag{2.75}$$

and, by replacing α with $1 - \alpha$,

$$\left|\langle T_j x, y\rangle\right|^2 \le \left\langle |T_j|^{2(1-\alpha)} x, x\right\rangle \left\langle |T_j^*|^{2\alpha} y, y\right\rangle, \tag{2.76}$$

for any $x, y \in H$.

By Hölder-McCarthy inequalities $\langle P^r x, x \rangle \leq \langle Px, x \rangle^r$, that holds for the positive operator P, for $r \in (0, 1)$ and $x \in H$ with $\|x\| = 1$ we also have

$$\left\langle |T_j|^{2\alpha} x, x \right\rangle \left\langle |T_j^*|^{2(1-\alpha)} y, y \right\rangle \leq \left\langle |T_j|^2 x, x \right\rangle^\alpha \left\langle |T_j^*|^2 y, y \right\rangle^{1-\alpha} \tag{2.77}$$

and

$$\left\langle |T_j|^{2(1-\alpha)} x, x \right\rangle \left\langle |T_j^*|^{2\alpha} y, y \right\rangle \leq \left\langle |T_j|^2 x, x \right\rangle^{1-\alpha} \left\langle |T_j^*|^2 y, y \right\rangle^\alpha \tag{2.78}$$

for any $x, y \in H$ with $\|x\| = \|y\| = 1$, $j \in \{1, ..., n\}$ and $\alpha \in (0, 1)$.

If we add (2.75) with (2.76) and make use of (2.77) and (2.78), we deduce

$$2 \left| \langle T_j x, y \rangle \right|^2 \leq \left\langle |T_j|^2 x, x \right\rangle^\alpha \left\langle |T_j^*|^2 y, y \right\rangle^{1-\alpha} + \left\langle |T_j^*|^2 y, y \right\rangle^\alpha \left\langle |T_j|^2 x, x \right\rangle^{1-\alpha} \tag{2.79}$$

for any $x, y \in H$ with $\|x\| = \|y\| = 1$, $j \in \{1, ..., n\}$ and $\alpha \in (0, 1)$.

Now, if we multiply (2.79) with $p_j \geq 0$, sum over j from 1 to n we get

$$2 \sum_{j=1}^n p_j \left| \langle T_j x, y \rangle \right|^2 \leq \sum_{j=1}^n p_j \left\langle |T_j|^2 x, x \right\rangle^\alpha \left\langle |T_j^*|^2 y, y \right\rangle^{1-\alpha} \tag{2.80}$$

$$+ \sum_{j=1}^n p_j \left\langle |T_j^*|^2 y, y \right\rangle^\alpha \left\langle |T_j|^2 x, x \right\rangle^{1-\alpha}$$

for any $x, y \in H$ with $\|x\| = \|y\| = 1$ and $\alpha \in (0, 1)$.

Since $\left\langle |T_j|^2 x, x \right\rangle = \|T_j x\|^2$ and $\left\langle |T_j^*|^2 y, y \right\rangle = \|T_j^* y\|^2$, $j \in \{1, ..., n\}$, then we get from (2.80) the first inequality in (2.74).

Now, on making use of the weighted Hölder discrete inequality

$$\sum_{j=1}^n p_j a_j b_j \leq \left(\sum_{j=1}^n p_j a_j^p \right)^{1/p} \left(\sum_{j=1}^n p_j b_j^q \right)^{1/q}, \, p, q > 1, \frac{1}{p} + \frac{1}{q} = 1,$$

where $(a_1, ..., a_n), (b_1, ..., b_n) \in \mathbb{R}_+^n$, we also have

$$\sum_{j=1}^n p_j \|T_j x\|^{2\alpha} \|T_j^* y\|^{2(1-\alpha)} \leq \left(\sum_{j=1}^n p_j \|T_j x\|^2 \right)^\alpha \left(\sum_{j=1}^n p_j \|T_j^* y\|^2 \right)^{1-\alpha}$$

and

$$\sum_{j=1}^{n} p_j \left\| T_j^* y \right\|^{2\alpha} \left\| T_j x \right\|^{2(1-\alpha)} \le \left(\sum_{j=1}^{n} p_j \left\| T_j^* y \right\|^2 \right)^{\alpha} \left(\sum_{j=1}^{n} p_j \left\| T_j x \right\|^2 \right)^{1-\alpha}.$$

Summing these two inequalities we deduce the second inequality in (2.74).

Finally, on utilizing the Hölder's inequality

$$ab + cd \le \left(a^p + c^p \right)^{1/p} \left(b^q + d^q \right)^{1/q}, a, b, c, d \ge 0$$

where $p > 1$ and $\frac{1}{p} + \frac{1}{q} = 1$, we have

$$\left(\sum_{j=1}^{n} p_j \left\| T_j x \right\|^2 \right)^{\alpha} \left(\sum_{j=1}^{n} p_j \left\| T_j^* y \right\|^2 \right)^{1-\alpha} + \left(\sum_{j=1}^{n} p_j \left\| T_j^* y \right\|^2 \right)^{\alpha} \left(\sum_{j=1}^{n} p_j \left\| T_j x \right\|^2 \right)^{1-\alpha}$$

$$\le \left(\sum_{j=1}^{n} p_j \left\| T_j x \right\|^2 + \sum_{j=1}^{n} p_j \left\| T_j^* y \right\|^2 \right)^{\alpha} \left(\sum_{j=1}^{n} p_j \left\| T_j x \right\|^2 + \sum_{j=1}^{n} p_j \left\| T_j^* y \right\|^2 \right)^{1-\alpha}$$

$$= \sum_{j=1}^{n} p_j \left\| T_j x \right\|^2 + \sum_{j=1}^{n} p_j \left\| T_j^* y \right\|^2.$$

and the proof is concluded.   ∎

Remark 2.25 For $\alpha = \frac{1}{2}$ we get from (2.74) that

$$\sum_{j=1}^{n} p_j \left| \langle T_j x, y \rangle \right|^2 \tag{2.81}$$

$$\le \sum_{j=1}^{n} p_j \left\| T_j x \right\| \left\| T_j^* y \right\| \le \left(\sum_{j=1}^{n} p_j \left\| T_j x \right\|^2 \right)^{1/2} \left(\sum_{j=1}^{n} p_j \left\| T_j^* y \right\|^2 \right)^{1/2}$$

$$\le \frac{1}{2} \sum_{j=1}^{n} p_j \left(\left\| T_j x \right\|^2 + \left\| T_j^* y \right\|^2 \right)$$

for any $x, y \in H$ with $\|x\| = \|y\| = 1$.

2.6 Inequalities for Functions of Normal Operators

Now, by the help of power series $f(z) = \sum_{n=0}^{\infty} a_n z^n$ we can naturally construct another power series which will have as coefficients the absolute values of the coefficient of the original series, namely, $f_A(z) := \sum_{n=0}^{\infty} |a_n| z^n$. It is obvious that this new power series will have the same radius of convergence as the original series. We also notice that if all coefficients $a_n \geq 0$, then $f_A = f$.

The following result is a functional inequality for normal operators that can be obtained from (2.64).

Theorem 2.26 (Dragomir et al. [9]) *Let* $f(z) = \sum_{n=0}^{\infty} a_n z^n$ *be a function defined by power series with complex coefficients and convergent on the open disk* $D(0, R) \subset \mathbb{C}$, $R > 0$. *If* N *is a normal operator on the Hilbert space* H *and for* $\alpha \in (0, 1)$ *we have that* $\|N\|^{2\alpha}$, $\|N\|^{2(1-\alpha)} < R$, *then we have the inequalities*

$$|\langle f(N) x, y \rangle| \leq \frac{1}{2} \left\langle \left[f_A \left(|N|^{2\alpha} \right) + f_A \left(|N|^{2(1-\alpha)} \right) \right] x, x \right\rangle^{1/2} \qquad (2.82)$$
$$\times \left\langle \left[f_A \left(|N|^{2\alpha} \right) + f_A \left(|N|^{2(1-\alpha)} \right) \right] y, y \right\rangle^{1/2}$$

for any $x, y \in H$.
 In particular, if $\|N\| < R$, *then*

$$|\langle f(N) x, y \rangle| \leq \langle f_A (|N|) x, x \rangle^{1/2} \langle f_A (|N|) y, y \rangle^{1/2} \qquad (2.83)$$

for any $x, y \in H$.

Proof If N is a normal operator, then for any $j \in \mathbb{N}$ we have that

$$\left| N^j \right|^2 = \left(N^* N \right)^j = |N|^{2j}.$$

Now, utilizing the inequality (2.75) we can write that

$$\left| \left\langle \sum_{j=0}^{n} a_j N^j x, y \right\rangle \right| \qquad (2.84)$$
$$\leq \sum_{j=0}^{n} |a_j| \left| \langle N^j x, y \rangle \right|$$
$$\leq \left\langle \sum_{j=0}^{n} |a_j| \left(\frac{|N|^{2j\alpha} + |N|^{2j(1-\alpha)}}{2} \right) x, x \right\rangle^{1/2}$$
$$\times \left\langle \sum_{j=0}^{n} |a_j| \left(\frac{|N|^{2j\alpha} + |N|^{2j(1-\alpha)}}{2} \right) y, y \right\rangle^{1/2}$$

for any $x, y \in H$ and $n \in \mathbb{N}$.

Since $\|N\|^{2\alpha}$, $\|N\|^{2(1-\alpha)} < R$, then it follows that the series $\sum_{j=0}^{\infty} |a_j| \left(|N|^{2\alpha}\right)^j$ and

$\sum_{j=0}^{\infty} |a_j| \left(|N|^{2(1-\alpha)}\right)^j$ are absolute convergent in $\mathcal{B}(H)$, and by taking the limit over $n \to \infty$ in (2.84) we deduce the desired result (2.82). ∎

Remark 2.27 With the assumptions in Theorem 2.26, if we take the supremum over $y \in H$, $\|y\| = 1$, then we get the vector inequality

$$\|f(N)x\| \leq \frac{1}{2} \left\langle \left[f_A \left(|N|^{2\alpha}\right) + f_A \left(|N|^{2(1-\alpha)}\right) \right] x, x \right\rangle^{1/2} \qquad (2.85)$$
$$\times \left\| f_A \left(|N|^{2\alpha}\right) + f_A \left(|N|^{2(1-\alpha)}\right) \right\|$$

for any $x \in H$, which in its turn produces the norm inequality

$$\|f(N)\| \leq \frac{1}{2} \left\| f_A \left(|N|^{2\alpha}\right) + f_A \left(|N|^{2(1-\alpha)}\right) \right\| \qquad (2.86)$$

for any $\alpha \in [0, 1]$.

Moreover, if we take $y = x$ in (2.82), then we have

$$|\langle f(N)x, x \rangle| \leq \frac{1}{2} \left\langle \left[f_A \left(|N|^{2\alpha}\right) + f_A \left(|N|^{2(1-\alpha)}\right) \right] x, x \right\rangle \qquad (2.87)$$

for any $x \in H$, which, by taking the supremum over $x \in H$, $\|x\| = 1$ generates the numerical radius inequality

$$w(f(N)) \leq \frac{1}{2} w \left[f_A \left(|N|^{2\alpha}\right) + f_A \left(|N|^{2(1-\alpha)}\right) \right] \qquad (2.88)$$

for any $\alpha \in [0, 1]$.

We can state the following particular vector inequalities:

$$\left| \left\langle \ln (1_H + N)^{-1} x, y \right\rangle \right| \qquad (2.89)$$
$$\leq \frac{1}{2} \left\langle \left[\ln \left(1_H - |N|^{2\alpha} \right)^{-1} + \ln \left(1_H - |N|^{2(1-\alpha)} \right)^{-1} \right] x, x \right\rangle^{1/2}$$
$$\times \left\langle \left[\ln \left(1_H - |N|^{2\alpha} \right)^{-1} + \ln \left(1_H - |N|^{2(1-\alpha)} \right)^{-1} \right] y, y \right\rangle^{1/2},$$

and

$$\left|\left\langle (1_H + N)^{-1} x, y \right\rangle\right| \tag{2.90}$$

$$\leq \frac{1}{2} \left\langle \left[\left(1_H - |N|^{2\alpha}\right)^{-1} + \left(1_H - |N|^{2(1-\alpha)}\right)^{-1} \right] x, x \right\rangle^{1/2}$$

$$\times \left\langle \left[\ln \left(1_H - |N|^{2\alpha}\right)^{-1} + \ln \left(1_H - |N|^{2(1-\alpha)}\right)^{-1} \right] y, y \right\rangle^{1/2},$$

for any $x, y \in H$ and $\|N\| < 1$.

We also have the inequalities

$$\left|\left\langle \sin (N) x, y \right\rangle\right| \leq \frac{1}{2} \left\langle \left[\sinh \left(|N|^{2\alpha}\right) + \sinh \left(|N|^{2(1-\alpha)}\right) \right] x, x \right\rangle^{1/2} \tag{2.91}$$

$$\times \left\langle \left[\sinh \left(|N|^{2\alpha}\right) + \sinh \left(|N|^{2(1-\alpha)}\right) \right] y, y \right\rangle^{1/2}$$

and

$$\left|\left\langle \cos (N) x, y \right\rangle\right| \leq \frac{1}{2} \left\langle \left[\cosh \left(|N|^{2\alpha}\right) + \cosh \left(|N|^{2(1-\alpha)}\right) \right] x, x \right\rangle^{1/2} \tag{2.92}$$

$$\times \left\langle \left[\cosh \left(|N|^{2\alpha}\right) + \cosh \left(|N|^{2(1-\alpha)}\right) \right] y, y \right\rangle^{1/2}$$

for any $x, y \in H$ and N a normal operator.

If we utilize functions as power series representations with nonnegative coefficients, then we can state the following vector inequalities:

$$\left|\left\langle \exp (N) x, y \right\rangle\right| \leq \frac{1}{2} \left\langle \left[\exp \left(|N|^{2\alpha}\right) + \exp \left(|N|^{2(1-\alpha)}\right) \right] x, x \right\rangle^{1/2} \tag{2.93}$$

$$\times \left\langle \left[\exp \left(|N|^{2\alpha}\right) + \exp \left(|N|^{2(1-\alpha)}\right) \right] y, y \right\rangle^{1/2}$$

for any $x, y \in H$ and N a normal operator.

If $\|N\| < 1$, then we also have the inequalities

$$\left|\left\langle \ln \left(\frac{1_H + N}{1_H - N} \right) x, y \right\rangle\right| \tag{2.94}$$

$$\leq \frac{1}{2} \left\langle \left[\ln \left(\frac{1_H + |N|^{2\alpha}}{1_H - |N|^{2\alpha}} \right) + \ln \left(\frac{1_H + |N|^{2(1-\alpha)}}{1_H - |N|^{2(1-\alpha)}} \right) \right] x, x \right\rangle^{1/2}$$

$$\times \left\langle \left[\ln \left(\frac{1_H + |N|^{2\alpha}}{1_H - |N|^{2\alpha}} \right) + \ln \left(\frac{1_H + |N|^{2(1-\alpha)}}{1_H - |N|^{2(1-\alpha)}} \right) \right] y, y \right\rangle^{1/2}$$

$$\left|\langle \tanh^{-1}(N)\, x, y\rangle\right| \tag{2.95}$$

$$\leq \frac{1}{2}\left\langle\left[\tanh^{-1}\left(|N|^{2\alpha}\right)+\tanh^{-1}\left(|N|^{2(1-\alpha)}\right)\right]x,x\right\rangle^{1/2}$$

$$\times\left\langle\left[\tanh^{-1}\left(|N|^{2\alpha}\right)+\tanh^{-1}\left(|N|^{2(1-\alpha)}\right)\right]y,y\right\rangle^{1/2}$$

and

$$\left|\langle {}_2F_1\left(\alpha,\beta,\gamma,N\right)x,y\rangle\right| \tag{2.96}$$

$$\leq \frac{1}{2}\left\langle\left[{}_2F_1\left(\alpha,\beta,\gamma,|N|^{2\alpha}\right)+{}_2F_1\left(\alpha,\beta,\gamma,|N|^{2(1-\alpha)}\right)\right]x,x\right\rangle^{1/2}$$

$$\times\left\langle\left[{}_2F_1\left(\alpha,\beta,\gamma,|N|^{2\alpha}\right)+{}_2F_1\left(\alpha,\beta,\gamma,|N|^{2(1-\alpha)}\right)\right]y,y\right\rangle^{1/2}$$

for any $x, y \in H$.

From a different perspective, we also have:

Theorem 2.28 (Dragomir et al. [9]) *With the assumption of Theorem 2.26 and if N is a normal operator on the Hilbert space H and $z \in \mathbb{C}$ such that $\|N\|^2$, $|z|^2 < R$, then we have the inequalities*

$$|\langle f(zN)x,y\rangle|^2 \leq \frac{1}{2}f_A\left(|z|^2\right)\left[\langle f_A\left(|N|^2\right)x,x\rangle^{\alpha}\langle f_A\left(|N|^2\right)y,y\rangle^{1-\alpha}\right. \tag{2.97}$$

$$\left.+\langle f_A\left(|N|^2\right)x,x\rangle^{1-\alpha}\langle f_A\left(|N|^2\right)y,y\rangle^{\alpha}\right]$$

$$\leq \frac{1}{2}f_A\left(|z|^2\right)\left(\langle f_A\left(|N|^2\right)x,x\rangle+\langle f_A\left(|N|^2\right)y,y\rangle\right)$$

for any $x, y \in H$ with $\|x\| = \|y\| = 1$ and $\alpha \in [0,1]$.

In particular, for $\alpha = \frac{1}{2}$ we have

$$|\langle f(zN)x,y\rangle|^2 \leq f_A\left(|z|^2\right)\langle f_A\left(|N|^2\right)x,x\rangle^{1/2}\langle f_A\left(|N|^2\right)y,y\rangle^{1/2} \tag{2.98}$$

$$\leq \frac{1}{2}f_A\left(|z|^2\right)\left(\langle f_A\left(|N|^2\right)x,x\rangle+\langle f_A\left(|N|^2\right)y,y\rangle\right)$$

for any $x, y \in H$ with $\|x\| = \|y\| = 1$.

Proof If we use the second and third inequality from (2.74) for powers of operators we have

$$\sum_{j=0}^{n}|a_j|\,|\langle N^j x, y\rangle|^2 \tag{2.99}$$

$$\leq \frac{1}{2}\left[\left(\sum_{j=0}^{n}|a_j|\,\|N^j x\|^2\right)^{\alpha}\left(\sum_{j=0}^{n}|a_j|\,\left\|(N^*)^j y\right\|^2\right)^{1-\alpha}\right]$$

$$+ \left(\sum_{j=0}^{n} |a_j| \, \|N^j x\|^2 \right)^{1-\alpha} \left(\sum_{j=0}^{n} |a_j| \, \|(N^*)^j y\|^2 \right)^{\alpha} \right]$$

$$\leq \frac{1}{2} \sum_{j=0}^{n} |a_j| \left(\|N^j x\|^2 + \|(N^*)^j y\|^2 \right)$$

for any $x, y \in H$ with $\|x\| = \|y\| = 1$ and $\alpha \in [0, 1]$.

Since N is a normal operator on the Hilbert space H, then

$$\|N^j x\|^2 = \left\langle |N^j|^2 x, x \right\rangle = \left\langle |N|^{2j} x, x \right\rangle$$

and

$$\|(N^*)^j y\|^2 = \left\langle |(N^*)^j|^2 y, y \right\rangle = \left\langle |N^*|^{2j} y, y \right\rangle = \left\langle |N|^{2j} y, y \right\rangle$$

for any $j \in \{0, ..., n\}$ and for any $x, y \in H$ with $\|x\| = \|y\| = 1$.

Then from (2.99) we have

$$\sum_{j=0}^{n} |a_j| \, |\langle N^j x, y \rangle|^2 \tag{2.100}$$

$$\leq \frac{1}{2} \left[\left(\left\langle \sum_{j=0}^{n} |a_j| \, |N|^{2j} x, x \right\rangle \right)^{\alpha} \left(\left\langle \sum_{j=0}^{n} |a_j| \, |N|^{2j} y, y \right\rangle \right)^{1-\alpha} \right.$$

$$\left. + \left(\left\langle \sum_{j=0}^{n} |a_j| \, |N|^{2j} x, x \right\rangle \right)^{1-\alpha} \left(\left\langle \sum_{j=0}^{n} |a_j| \, |N|^{2j} y, y \right\rangle \right)^{\alpha} \right]$$

$$\leq \frac{1}{2} \left(\left\langle \sum_{j=0}^{n} |a_j| \, |N|^{2j} x, x \right\rangle + \left\langle \sum_{j=0}^{n} |a_j| \, |N|^{2j} y, y \right\rangle \right)$$

for any $x, y \in H$ with $\|x\| = \|y\| = 1$ and $\alpha \in [0, 1]$.

By the weighted Cauchy–Buniakowski–Schwarz inequality we also have

$$\left| \left\langle \sum_{j=0}^{n} a_j z^j N^j x, y \right\rangle \right|^2 \leq \sum_{j=0}^{n} |a_j| \, |z|^{2j} \sum_{j=0}^{n} |a_j| \, |\langle N^j x, y \rangle|^2 \tag{2.101}$$

for any $x, y \in H$ with $\|x\| = \|y\| = 1$.

Now, since the series $\sum_{j=0}^{\infty} a_j z^j N^j$, $\sum_{j=0}^{\infty} |a_j| \, |z|^{2j}$, $\sum_{j=0}^{\infty} |a_j| \, |N|^{2j}$ are convergent, then by (2.100) and (2.101), on letting $n \to \infty$, we deduce the desired result (2.97). ∎

Similar inequalities for some particular functions of interest can be stated. However, the details are left to the interested reader.

2.7 Applications for the Euclidian Norm

We can state now the following result:

Theorem 2.29 (Dragomir et al. [9]) *For any* $(T_1, \ldots, T_n) \in B^{(n)}(H)$ *we have*

$$\|(T_1, \ldots, T_n)\|_e^2 \leq \frac{1}{2} \left[\left(\left\| \sum_{j=1}^n |T_j|^2 \right\| \right)^\alpha \left(\left\| \sum_{j=1}^n |T_j^*|^2 \right\| \right)^{1-\alpha} \right. \tag{2.102}$$
$$\left. + \left(\left\| \sum_{j=1}^n |T_j|^2 \right\| \right)^{1-\alpha} \left(\left\| \sum_{j=1}^n |T_j^*|^2 \right\| \right)^\alpha \right]$$
$$\leq \frac{1}{2} \left[\left\| \sum_{j=1}^n |T_j|^2 \right\| + \left\| \sum_{j=1}^n |T_j^*|^2 \right\| \right]$$

and

$$w_e^2(T_1, \ldots, T_n) \tag{2.103}$$
$$\leq \frac{1}{2} \left[\sup_{\|x\|=1} \left\{ \left(\left\langle \sum_{j=1}^n |T_j|^2 x, x \right\rangle \right)^\alpha \left(\left\langle \sum_{j=1}^n |T_j^*|^2 x, x \right\rangle \right)^{1-\alpha} \right\} \right.$$
$$\left. + \sup_{\|x\|=1} \left\{ \left(\left\langle \sum_{j=1}^n |T_j|^2 x, x \right\rangle \right)^{1-\alpha} \left(\left\langle \sum_{j=1}^n |T_j^*|^2 x, x \right\rangle \right)^\alpha \right\} \right]$$
$$\leq \frac{1}{2} \left[\left(\left\| \sum_{j=1}^n |T_j|^2 \right\| \right)^\alpha \left(\left\| \sum_{j=1}^n |T_j^*|^2 \right\| \right)^{1-\alpha} \right.$$
$$\left. + \left(\left\| \sum_{j=1}^n |T_j|^2 \right\| \right)^{1-\alpha} \left(\left\| \sum_{j=1}^n |T_j^*|^2 \right\| \right)^\alpha \right]$$

for any $\alpha \in [0, 1]$.

Proof We have from the second inequality in (2.74)

$$\sum_{j=1}^{n} |\langle T_j x, y\rangle|^2 \le \frac{1}{2}\left[\left(\left\langle\sum_{j=1}^{n}|T_j|^2 x, x\right\rangle\right)^{\alpha}\left(\left\langle\sum_{j=1}^{n}|T_j^*|^2 y, y\right\rangle\right)^{1-\alpha}\right.\tag{2.104}$$

$$\left.+\left(\left\langle\sum_{j=1}^{n}|T_j|^2 x, x\right\rangle\right)^{1-\alpha}\left(\left\langle\sum_{j=1}^{n}|T_j^*|^2 y, y\right\rangle\right)^{\alpha}\right]$$

for any $x, y \in H$ with $\|x\| = \|y\| = 1$ and $\alpha \in [0, 1]$.

Taking the supremum over $\|x\| = \|y\| = 1$ we have

$$\|(T_1, \ldots, T_n)\|_e^2$$

$$\le \frac{1}{2}\left[\left(\sup_{\|x\|=1}\left\langle\sum_{j=1}^{n}|T_j|^2 x, x\right\rangle\right)^{\alpha}\left(\sup_{\|y\|=1}\left\langle\sum_{j=1}^{n}|T_j^*|^2 y, y\right\rangle\right)^{1-\alpha}\right.$$

$$\left.+\left(\sup_{\|x\|=1}\left\langle\sum_{j=1}^{n}|T_j|^2 x, x\right\rangle\right)^{1-\alpha}\left(\sup_{\|y\|=1}\left\langle\sum_{j=1}^{n}|T_j^*|^2 y, y\right\rangle\right)^{\alpha}\right]$$

$$= \frac{1}{2}\left[\left(\left\|\sum_{j=1}^{n}|T_j|^2\right\|\right)^{\alpha}\left(\left\|\sum_{j=1}^{n}|T_j^*|^2\right\|\right)^{1-\alpha}\right.$$

$$\left.+\left(\left\|\sum_{j=1}^{n}|T_j|^2\right\|\right)^{1-\alpha}\left(\left\|\sum_{j=1}^{n}|T_j^*|^2\right\|\right)^{\alpha}\right],$$

which proves the first part of (2.102).

The second part follows by the elementary inequality

$$a^{\alpha}b^{1-\alpha} \le \alpha a + (1 - \alpha)b$$

for $a, b \ge 0$ and $\alpha \in [0, 1]$.

The inequality (2.103) follows from (2.104) by taking $y = x$ and then the supremum over $\|x\| = 1$.   ∎

2.8 Applications for s-1-Norm and s-1-Numerical Radius

We can state the following result:

Theorem 2.30 (Dragomir et al. [9]) *For any* $(T_1, \ldots, T_n) \in B^{(n)}(H)$ *we have*

$$\|(T_1, \ldots, T_n)\|_{s,1} \leq \left\| \sum_{j=1}^{n} \left(\frac{|T_j|^{2\alpha} + |T_j|^{2(1-\alpha)}}{2} \right) \right\|^{1/2} \tag{2.105}$$

$$\times \left\| \sum_{j=1}^{n} \left(\frac{\left|T_j^*\right|^{2\alpha} + \left|T_j^*\right|^{2(1-\alpha)}}{2} \right) \right\|^{1/2}$$

$$\leq \frac{1}{2} \left[\left\| \sum_{j=1}^{n} \left(\frac{|T_j|^{2\alpha} + |T_j|^{2(1-\alpha)}}{2} \right) \right\| \right.$$

$$\left. + \left\| \sum_{j=1}^{n} \left(\frac{\left|T_j^*\right|^{2\alpha} + \left|T_j^*\right|^{2(1-\alpha)}}{2} \right) \right\| \right]$$

and

$$w_{s,1}(T_1, \ldots, T_n) \tag{2.106}$$

$$\leq \left\| \sum_{j=1}^{n} \left(\frac{|T_j|^{2\alpha} + |T_j|^{2(1-\alpha)} + \left|T_j^*\right|^{2\alpha} + \left|T_j^*\right|^{2(1-\alpha)}}{4} \right) \right\|.$$

Proof From (2.64) we have

$$\sum_{j=1}^{n} |\langle T_j x, y \rangle| \leq \left\langle \sum_{j=1}^{n} \left(\frac{|T_j|^{2\alpha} + |T_j|^{2(1-\alpha)}}{2} \right) x, x \right\rangle^{1/2} \tag{2.107}$$

$$\times \left\langle \sum_{j=1}^{n} \left(\frac{\left|T_j^*\right|^{2\alpha} + \left|T_j^*\right|^{2(1-\alpha)}}{2} \right) y, y \right\rangle^{1/2}$$

for any $x, y \in H$.

Taking the supremum over $\|y\| = 1$, $\|x\| = 1$ in (2.107) we have

$$\|(T_1, \ldots, T_n)\|_{s,1} \leq \left[\sup_{\|x\|=1} \left\langle \sum_{j=1}^{n} \left(\frac{|T_j|^{2\alpha} + |T_j|^{2(1-\alpha)}}{2} \right) x, x \right\rangle \right]^{1/2}$$

$$\times \left[\sup_{\|y\|=1} \left\langle \sum_{j=1}^{n} \left(\frac{\left|T_j^*\right|^{2\alpha} + \left|T_j^*\right|^{2(1-\alpha)}}{2} \right) y, y \right\rangle \right]^{1/2}$$

$$= \left\| \sum_{j=1}^{n} \left(\frac{|T_j|^{2\alpha} + |T_j|^{2(1-\alpha)}}{2} \right) \right\|^{1/2}$$

$$\times \left\| \sum_{j=1}^{n} \left(\frac{\left|T_j^*\right|^{2\alpha} + \left|T_j^*\right|^{2(1-\alpha)}}{2} \right) \right\|^{1/2}$$

and the first inequality (2.105) is proved.

The second part follows by the arithmetic mean-geometric mean inequality.

Now, if we take $y = x$ in (2.107), then we get

$$\sum_{j=1}^{n} |\langle T_j x, x \rangle| \le \left\langle \sum_{j=1}^{n} \left(\frac{|T_j|^{2\alpha} + |T_j|^{2(1-\alpha)}}{2} \right) x, x \right\rangle^{1/2}$$

$$\times \left\langle \sum_{j=1}^{n} \left(\frac{\left|T_j^*\right|^{2\alpha} + \left|T_j^*\right|^{2(1-\alpha)}}{2} \right) x, x \right\rangle^{1/2}$$

$$\le \frac{1}{2} \left\langle \sum_{j=1}^{n} \left(\frac{|T_j|^{2\alpha} + |T_j|^{2(1-\alpha)} + \left|T_j^*\right|^{2\alpha} + \left|T_j^*\right|^{2(1-\alpha)}}{2} \right) x, x \right\rangle.$$

Taking the supremum over $\|x\| = 1$ we deduce the desired result (2.106). ∎

Remark 2.31 If we take $\alpha = \frac{1}{2}$ in the first inequality in (2.105), then we deduce

$$\|(T_1, \dots, T_n)\|_{s,1} \le \left\| \sum_{j=1}^{n} |T_j| \right\|^{1/2} \left\| \sum_{j=1}^{n} |T_j^*| \right\|^{1/2} \tag{2.108}$$

from where we get the following refinement of the generalized triangle inequality

$$\left\| \sum_{j=1}^{n} T_j \right\| \le \|(T_1, \dots, T_n)\|_{s,1} \le \left\| \sum_{j=1}^{n} |T_j| \right\|^{1/2} \left\| \sum_{j=1}^{n} |T_j^*| \right\|^{1/2}$$

$$\le \frac{1}{2} \left[\left\| \sum_{j=1}^{n} |T_j| \right\| + \left\| \sum_{j=1}^{n} |T_j^*| \right\| \right] \le \sum_{j=1}^{n} \|T_j\|.$$

From (2.106) we also have for $\alpha = \frac{1}{2}$ that

$$w_{s,1}(T_1, \ldots, T_n) \leq \left\| \sum_{j=1}^{n} \left(\frac{|T_j| + |T_j^*|}{2} \right) \right\|. \tag{2.109}$$

2.9 Other Additive Inequalities

The following result holds:

Theorem 2.32 (Dragomir et al. [10]) *Let* $(T_1, \ldots, T_n) \in \mathcal{B}^{(n)}(H)$ *be an n-tuple of bounded linear operators on the Hilbert space* $(H; \langle \cdot, \cdot \rangle)$ *and* $(p_1, \ldots, p_n) \in \mathbb{R}_+^{*n}$ *an n-tuple of nonnegative weights not all of them equal to zero, then*

$$\left| \left\langle \sum_{j=1}^{n} p_j \left(\frac{T_j + T_j^*}{2} \right) x, y \right\rangle \right| \leq \sum_{j=1}^{n} p_j \left| \left\langle \frac{T_j + T_j^*}{2} x, y \right\rangle \right| \tag{2.110}$$

$$\leq \sum_{j=1}^{n} p_j \left[\frac{|\langle T_j x, y \rangle| + |\langle T_j^* x, y \rangle|}{2} \right]$$

$$\leq \left\langle \sum_{j=1}^{n} p_j \left[\frac{|T_j|^{2\alpha} + |T_j^*|^{2(1-\alpha)}}{2} \right] x, x \right\rangle^{1/2}$$

$$\times \left\langle \sum_{j=1}^{n} p_j \left[\frac{|T_j|^{2\alpha} + |T_j^*|^{2(1-\alpha)}}{2} \right] y, y \right\rangle^{1/2}$$

for any $\alpha \in [0, 1]$ *and, in particular, for* $\alpha = \frac{1}{2}$

$$\left| \left\langle \sum_{j=1}^{n} p_j \left(\frac{T_j + T_j^*}{2} \right) x, y \right\rangle \right| \leq \sum_{j=1}^{n} p_j \left| \left\langle \frac{T_j + T_j^*}{2} x, y \right\rangle \right| \tag{2.111}$$

$$\leq \sum_{j=1}^{n} p_j \left[\frac{|\langle T_j x, y \rangle| + |\langle T_j^* x, y \rangle|}{2} \right]$$

$$\leq \left\langle \sum_{j=1}^{n} p_j \left[\frac{|T_j| + |T_j^*|}{2}\right] x, x \right\rangle^{1/2}$$

$$\times \left\langle \sum_{j=1}^{n} p_j \left[\frac{|T_j| + |T_j^*|}{2}\right] y, y \right\rangle^{1/2}$$

for any $x, y \in H$.

Proof The first two inequalities are obvious by the properties of the modulus.

Utilising Kato's inequality we have

$$|\langle T_j x, y \rangle| \leq \left\langle |T_j|^{2\alpha} x, x \right\rangle^{1/2} \left\langle |T_j^*|^{2(1-\alpha)} y, y \right\rangle^{1/2} \tag{2.112}$$

and, by replacing x with y we have

$$|\langle T_j y, x \rangle| \leq \left\langle |T_j|^{2\alpha} y, y \right\rangle^{1/2} \left\langle |T_j^*|^{2(1-\alpha)} x, x \right\rangle^{1/2}$$

i.e.,

$$|\langle T_j^* x, y \rangle| \leq \left\langle |T_j^*|^{2(1-\alpha)} x, x \right\rangle^{1/2} \left\langle |T_j|^{2\alpha} y, y \right\rangle^{1/2} \tag{2.113}$$

for any $j \in \{1, ..., n\}$ and $x, y \in H$.

Adding the inequalities (2.112) and (2.113) and utilizing the elementary inequality

$$ab + cd \leq \left(a^2 + c^2\right)^{1/2} \left(b^2 + d^2\right)^{1/2}, a, b, c, d \geq 0$$

we get

$$|\langle T_j x, y \rangle| + |\langle T_j^* x, y \rangle| \leq \left\langle |T_j|^{2\alpha} x, x \right\rangle^{1/2} \left\langle |T_j^*|^{2(1-\alpha)} y, y \right\rangle^{1/2} \tag{2.114}$$

$$+ \left\langle |T_j^*|^{2(1-\alpha)} x, x \right\rangle^{1/2} \left\langle |T_j|^{2\alpha} y, y \right\rangle^{1/2}$$

$$\leq \left\langle \left[|T_j|^{2\alpha} + |T_j^*|^{2(1-\alpha)}\right] x, x \right\rangle^{1/2}$$

$$\times \left\langle \left[|T_j|^{2\alpha} + |T_j^*|^{2(1-\alpha)}\right] y, y \right\rangle^{1/2}$$

for any $j \in \{1, ..., n\}$ and $x, y \in H$.

Multiplying the inequalities (2.114) by $p_j \geq 0$ and then summing over j from 1 to n and utilizing the weighted Cauchy–Buniakowski–Schwarz inequality we have

$$\sum_{j=1}^{n} p_j \left[|\langle T_j x, y \rangle| + |\langle T_j^* x, y \rangle|\right] \tag{2.115}$$

$$\leq \sum_{j=1}^{n} p_j \left\langle \left[|T_j|^{2\alpha} + |T_j^*|^{2(1-\alpha)} \right] x, x \right\rangle^{1/2} \left\langle \left[|T_j|^{2\alpha} + |T_j^*|^{2(1-\alpha)} \right] y, y \right\rangle^{1/2}$$

$$\leq \left\langle \sum_{j=1}^{n} p_j \left[|T_j|^{2\alpha} + |T_j^*|^{2(1-\alpha)} \right] x, x \right\rangle^{1/2}$$

$$\times \left\langle \sum_{j=1}^{n} p_j \left[|T_j|^{2\alpha} + |T_j^*|^{2(1-\alpha)} \right] y, y \right\rangle^{1/2}$$

for $x, y \in H$, which is equivalent with the third inequality in (2.110). ∎

Remark 2.33 The particular case $y = x$ is of interest for providing numerical radii inequalities and can be stated as:

$$\left| \left\langle \sum_{j=1}^{n} p_j \left(\frac{T_j + T_j^*}{2} \right) x, x \right\rangle \right| \leq \sum_{j=1}^{n} p_j \left| \left\langle \frac{T_j + T_j^*}{2} x, x \right\rangle \right| \tag{2.116}$$

$$\leq \sum_{j=1}^{n} p_j \left| \langle T_j x, x \rangle \right|$$

$$\leq \left\langle \sum_{j=1}^{n} p_j \left[\frac{|T_j|^{2\alpha} + \left| T_j^* \right|^{2(1-\alpha)}}{2} \right] x, x \right\rangle$$

for any $\alpha \in [0, 1]$ and, for $\alpha = \frac{1}{2}$,

$$\left| \left\langle \sum_{j=1}^{n} p_j \left(\frac{T_j + T_j^*}{2} \right) x, x \right\rangle \right| \leq \sum_{j=1}^{n} p_j \left| \left\langle \frac{T_j + T_j^*}{2} x, x \right\rangle \right| \tag{2.117}$$

$$\leq \sum_{j=1}^{n} p_j \left| \langle T_j x, x \rangle \right|$$

$$\leq \left\langle \sum_{j=1}^{n} p_j \left[\frac{|T_j| + \left| T_j^* \right|}{2} \right] x, x \right\rangle$$

for any $x \in H$.

The case of unitary vectors provides more refinements as follows:

Remark 2.34 With the assumptions in Theorem 2.32, we have

$$\sum_{j=1}^{n} p_j \left[\frac{\left| \langle T_j x, y \rangle \right| + \left| \langle T_j^* x, y \rangle \right|}{2} \right]$$ (2.118)

$$\leq \left\langle \sum_{j=1}^{n} p_j \left[\frac{\left| T_j \right|^{2\alpha} + \left| T_j^* \right|^{2(1-\alpha)}}{2} \right] x, x \right\rangle^{1/2}$$

$$\times \left\langle \sum_{j=1}^{n} p_j \left[\frac{\left| T_j \right|^{2\alpha} + \left| T_j^* \right|^{2(1-\alpha)}}{2} \right] y, y \right\rangle^{1/2}$$

$$\leq \left\langle \left(\sum_{j=1}^{n} p_j \left[\frac{\left| T_j \right|^{2\alpha} + \left| T_j^* \right|^{2(1-\alpha)}}{2} \right] \right)^{1/2} x, x \right\rangle$$

$$\times \left\langle \left(\sum_{j=1}^{n} p_j \left[\frac{\left| T_j \right|^{2\alpha} + \left| T_j^* \right|^{2(1-\alpha)}}{2} \right] \right)^{1/2} y, y \right\rangle$$

$$\leq \frac{1}{2} \left[\left\langle \left(\sum_{j=1}^{n} p_j \left[\frac{\left| T_j \right|^{2\alpha} + \left| T_j^* \right|^{2(1-\alpha)}}{2} \right] \right)^{1/2} x, x \right\rangle^2 \right.$$

$$\left. + \left\langle \left(\sum_{j=1}^{n} p_j \left[\frac{\left| T_j \right|^{2\alpha} + \left| T_j^* \right|^{2(1-\alpha)}}{2} \right] \right)^{1/2} y, y \right\rangle^2 \right]$$

$$\leq \frac{1}{2} \left[\left\langle \sum_{j=1}^{n} p_j \left[\frac{\left| T_j \right|^{2\alpha} + \left| T_j^* \right|^{2(1-\alpha)}}{2} \right] x, x \right\rangle \right.$$

$$\left. + \left\langle \sum_{j=1}^{n} p_j \left[\frac{\left| T_j \right|^{2\alpha} + \left| T_j^* \right|^{2(1-\alpha)}}{2} \right] y, y \right\rangle \right]$$

for any $\alpha \in [0, 1]$ and, in particular,

$$\sum_{j=1}^{n} p_j \left[\frac{|\langle T_j x, y \rangle| + |\langle T_j^* x, y \rangle|}{2} \right] \qquad (2.119)$$

$$\leq \left\langle \sum_{j=1}^{n} p_j \left[\frac{|T_j| + |T_j^*|}{2} \right] x, x \right\rangle^{1/2} \left\langle \sum_{j=1}^{n} p_j \left[\frac{|T_j| + |T_j^*|}{2} \right] y, y \right\rangle^{1/2}$$

$$\leq \left\langle \left(\sum_{j=1}^{n} p_j \left[\frac{|T_j| + |T_j^*|}{2} \right] \right)^{1/2} x, x \right\rangle \left\langle \left(\sum_{j=1}^{n} p_j \left[\frac{|T_j| + |T_j^*|}{2} \right] \right)^{1/2} y, y \right\rangle$$

$$\leq \frac{1}{2} \left[\left\langle \left(\sum_{j=1}^{n} p_j \left[\frac{|T_j| + |T_j^*|}{2} \right] \right)^{1/2} x, x \right\rangle^2 \right.$$

$$+ \left\langle \left(\sum_{j=1}^{n} p_j \left[\frac{|T_j| + |T_j^*|}{2} \right] \right)^{1/2} y, y \right\rangle^2 \right]$$

$$\leq \frac{1}{2} \left[\left\langle \sum_{j=1}^{n} p_j \left[\frac{|T_j| + |T_j^*|}{2} \right] x, x \right\rangle + \left\langle \sum_{j=1}^{n} p_j \left[\frac{|T_j| + |T_j^*|}{2} \right] y, y \right\rangle \right]$$

for any $x, y \in H$ with $\|x\| = \|y\| = 1$.

The proofs follow by utilizing the Hölder-McCarthy inequalities $\langle P^r x, x \rangle \leq \langle P x, x \rangle^r$ and $\langle P x, x \rangle^s \leq \langle P^s x, x \rangle$ that hold for the positive operator P, for $r \in (0, 1)$, $s \in [1, \infty)$ and $x \in H$ with $\|x\| = 1$. The details are omitted.

In order to employ the above result in obtaining some inequalities for functions of normal operators defined by power series, we need the following version of (2.110).

Remark 2.35 If we write the inequality (2.110) for the normal operators N_j, $j \in \{1, ..., n\}$ then we get

$$\left| \left\langle \sum_{j=1}^{n} p_j \left(\frac{N_j + N_j^*}{2} \right) x, y \right\rangle \right| \leq \sum_{j=1}^{n} p_j \left| \left\langle \frac{N_j + N_j^*}{2} x, y \right\rangle \right| \qquad (2.120)$$

$$\leq \sum_{j=1}^{n} p_j \left[\frac{|\langle N_j x, y \rangle| + |\langle N_j^* x, y \rangle|}{2} \right]$$

$$\leq \left\langle \sum_{j=1}^{n} p_j \left[\frac{|N_j|^{2\alpha} + |N_j|^{2(1-\alpha)}}{2} \right] x, x \right\rangle^{1/2}$$

$$\times \left\langle \sum_{j=1}^{n} p_j \left[\frac{|N_j|^{2\alpha} + |N_j|^{2(1-\alpha)}}{2} \right] y, y \right\rangle^{1/2}$$

for any $\alpha \in [0, 1]$ and, in particular, for $\alpha = \frac{1}{2}$

$$\left| \left\langle \sum_{j=1}^{n} p_j \left(\frac{N_j + N_j^*}{2} \right) x, y \right\rangle \right| \leq \sum_{j=1}^{n} p_j \left| \left\langle \frac{N_j + N_j^*}{2} x, y \right\rangle \right| \tag{2.121}$$

$$\leq \sum_{j=1}^{n} p_j \left[\frac{|\langle N_j x, y \rangle| + |\langle N_j^* x, y \rangle|}{2} \right]$$

$$\leq \left\langle \sum_{j=1}^{n} p_j |N_j| x, x \right\rangle^{1/2} \left\langle \sum_{j=1}^{n} p_j |N_j| y, y \right\rangle^{1/2}$$

for any $x, y \in H$.

The following results involving quadratics also holds:

Theorem 2.36 (Dragomir et al. [10]) *Let $(T_1, ..., T_n) \in \mathcal{B}^{(n)}(H)$ be an n-tuple of bounded linear operators on the Hilbert space $(H; \langle \cdot, \cdot \rangle)$ and $(p_1, ..., p_n) \in \mathbb{R}_+^{*n}$ an n-tuple of nonnegative weights not all of them equal to zero, then*

$$\sum_{j=1}^{n} p_j \left[|\langle T_j x, y \rangle|^2 + |\langle T_j^* x, y \rangle|^2 \right] \tag{2.122}$$

$$\leq \sum_{j=1}^{n} p_j \left[\|T_j x\|^{2\alpha} \|T_j^* y\|^{2(1-\alpha)} + \|T_j y\|^{2\alpha} \|T_j^* x\|^{2(1-\alpha)} \right]$$

$$\leq \left(\sum_{j=1}^{n} p_j \|T_j x\|^2 \right)^{\alpha} \left(\sum_{j=1}^{n} p_j \|T_j^* y\|^2 \right)^{1-\alpha}$$

$$+ \left(\sum_{j=1}^{n} p_j \|T_j y\|^2 \right)^{\alpha} \left(\sum_{j=1}^{n} p_j \|T_j^* x\|^2 \right)^{1-\alpha}$$

$$\leq \left(\sum_{j=1}^{n} p_j \left[\|T_j x\|^2 + \|T_j y\|^2 \right] \right)^{\alpha} \left(\sum_{j=1}^{n} p_j \left[\|T_j^* y\|^2 + \|T_j^* x\|^2 \right] \right)^{1-\alpha}$$

for any $x, y \in H$ with $\|x\| = \|y\| = 1$ and $\alpha \in [0, 1]$.

Proof We must prove the inequalities only in the case $\alpha \in (0, 1)$, since the case $\alpha = 0$ or $\alpha = 1$ follows directly from the corresponding case of Kato's inequality.

Utilising Kato's inequality we have

$$|\langle T_j x, y \rangle|^2 \leq \left\langle |T_j|^{2\alpha} x, x \right\rangle \left\langle |T_j^*|^{2(1-\alpha)} y, y \right\rangle \tag{2.123}$$

and, by replacing x with y we have

$$|\langle T_j^* x, y \rangle|^2 \leq \left\langle |T_j^*|^{2(1-\alpha)} x, x \right\rangle \left\langle |T_j|^{2\alpha} y, y \right\rangle \tag{2.124}$$

for any $j \in \{1, ..., n\}$ and $x, y \in H$.

By Hölder-McCarthy inequality $\langle P^r x, x \rangle \leq \langle Px, x \rangle^r$ for $r \in (0, 1)$ and $x \in H$ with $\|x\| = 1$ we also have

$$\left\langle |T_j|^{2\alpha} x, x \right\rangle \left\langle |T_j^*|^{2(1-\alpha)} y, y \right\rangle \leq \left\langle |T_j|^2 x, x \right\rangle^\alpha \left\langle |T_j^*|^2 y, y \right\rangle^{1-\alpha} \tag{2.125}$$

and

$$\left\langle |T_j^*|^{2(1-\alpha)} x, x \right\rangle \left\langle |T_j|^{2\alpha} y, y \right\rangle \leq \left\langle |T_j|^2 y, y \right\rangle^\alpha \left\langle |T_j^*|^2 x, x \right\rangle^{1-\alpha} \tag{2.126}$$

for any $j \in \{1, ..., n\}$ and $x, y \in H$ with $\|x\| = \|y\| = 1$.

We then obtain by summation

$$|\langle T_j x, y \rangle|^2 + |\langle T_j^* x, y \rangle|^2 \tag{2.127}$$
$$\leq \left\langle |T_j|^2 x, x \right\rangle^\alpha \left\langle |T_j^*|^2 y, y \right\rangle^{1-\alpha} + \left\langle |T_j|^2 y, y \right\rangle^\alpha \left\langle |T_j^*|^2 x, x \right\rangle^{1-\alpha}$$

for any $j \in \{1, ..., n\}$ and $x, y \in H$ with $\|x\| = \|y\| = 1$.

Now, if we multiply (2.127) with $p_j \geq 0$, sum over j from 1 to n we get

$$\sum_{j=1}^n p_j \left[|\langle T_j x, y \rangle|^2 + |\langle T_j^* x, y \rangle|^2 \right] \tag{2.128}$$

$$\leq \sum_{j=1}^n p_j \left\langle |T_j|^2 x, x \right\rangle^\alpha \left\langle |T_j^*|^2 y, y \right\rangle^{1-\alpha}$$

$$+ \sum_{j=1}^n p_j \left\langle |T_j|^2 y, y \right\rangle^\alpha \left\langle |T_j^*|^2 x, x \right\rangle^{1-\alpha}$$

for any $x, y \in H$ with $\|x\| = \|y\| = 1$ and $\alpha \in (0, 1)$.

Since $\left\langle \left|T_j\right|^2 x, x\right\rangle = \left\|T_j x\right\|^2, \left\langle \left|T_j^*\right|^2 y, y\right\rangle = \left\|T_j^* y\right\|^2, \left\langle \left|T_j\right|^2 y, y\right\rangle = \left\|T_j y\right\|^2$ and $\left\langle \left|T_j^*\right|^2 x, x\right\rangle = \left\|T_j^* x\right\|^2$ $j \in \{1, ..., n\}$, then we get from (2.128) the first part of (2.122).

Now, on making use of the weighted Hölder discrete inequality

$$\sum_{j=1}^n p_j a_j b_j \leq \left(\sum_{j=1}^n p_j a_j^p\right)^{1/p} \left(\sum_{j=1}^n p_j b_j^q\right)^{1/q}, p, q > 1, \frac{1}{p} + \frac{1}{q} = 1,$$

where $(a_1, ..., a_n), (b_1, ..., b_n) \in \mathbb{R}_+^n$, we also have

$$\sum_{j=1}^n p_j \left\|T_j x\right\|^{2\alpha} \left\|T_j^* y\right\|^{2(1-\alpha)} \leq \left(\sum_{j=1}^n p_j \left\|T_j x\right\|^2\right)^\alpha \left(\sum_{j=1}^n p_j \left\|T_j^* y\right\|^{2(1-\alpha)}\right)^{1-\alpha}$$

and

$$\sum_{j=1}^n p_j \left\|T_j y\right\|^{2\alpha} \left\|T_j^* x\right\|^{2(1-\alpha)} \leq \left(\sum_{j=1}^n p_j \left\|T_j y\right\|^2\right)^\alpha \left(\sum_{j=1}^n p_j \left\|T_j^* x\right\|^2\right)^{1-\alpha}.$$

Summing these two inequalities we deduce the second inequality in (2.122).

Finally, on utilizing the Hölder inequality

$$ab + cd \leq \left(a^p + c^p\right)^{1/p} \left(b^q + d^q\right)^{1/q}, a, b, c, d \geq 0$$

where $p > 1$ and $\frac{1}{p} + \frac{1}{q} = 1$, we have

$$\left(\sum_{j=1}^n p_j \left\|T_j x\right\|^2\right)^\alpha \left(\sum_{j=1}^n p_j \left\|T_j^* y\right\|^2\right)^{1-\alpha}$$

$$+ \left(\sum_{j=1}^n p_j \left\|T_j y\right\|^2\right)^\alpha \left(\sum_{j=1}^n p_j \left\|T_j^* x\right\|^2\right)^{1-\alpha}$$

$$\leq \left(\sum_{j=1}^n p_j \left\|T_j x\right\|^2 + \sum_{j=1}^n p_j \left\|T_j y\right\|^2\right)^\alpha \left(\sum_{j=1}^n p_j \left\|T_j^* y\right\|^2 + \sum_{j=1}^n p_j \left\|T_j^* x\right\|^2\right)^{1-\alpha}$$

and the proof is completed. ∎

Remark 2.37 Utilizing the elementary inequality for complex numbers

$$\left|\frac{z+w}{2}\right|^2 \leq \frac{|z|^2+|w|^2}{2}, z, w \in \mathbb{C}$$

we have

$$\sum_{j=1}^{n} p_j \left[\left|\left\langle\left(\frac{T_j+T_j^*}{2}\right)x, y\right\rangle\right|^2\right] \leq \sum_{j=1}^{n} p_j \left[\frac{|\langle T_j x, y\rangle|^2 + |\langle T_j^* x, y\rangle|^2}{2}\right] \quad (2.129)$$

and by the weighted arithmetic mean-geometric mean inequality

$$a^\alpha b^{1-\alpha} \leq \alpha a + (1-\alpha)b, a, b \geq 0, \alpha \in [0,1]$$

we also have

$$\left(\sum_{j=1}^{n} p_j \left[\|T_j x\|^2 + \|T_j y\|^2\right]\right)^\alpha \left(\sum_{j=1}^{n} p_j \left[\|T_j^* y\|^2 + \|T_j^* x\|^2\right]\right)^{1-\alpha} \quad (2.130)$$

$$\leq \alpha \sum_{j=1}^{n} p_j \left[\|T_j x\|^2 + \|T_j y\|^2\right] + (1-\alpha)\sum_{j=1}^{n} p_j \left[\|T_j^* y\|^2 + \|T_j^* x\|^2\right].$$

If we choose $\alpha = \frac{1}{2}$ and use (2.113), (2.129) and (2.130) we derive

$$\sum_{j=1}^{n} p_j \left[\left|\left\langle\left(\frac{T_j+T_j^*}{2}\right)x, y\right\rangle\right|^2\right] \quad (2.131)$$

$$\leq \sum_{j=1}^{n} p_j \left[\frac{|\langle T_j x, y\rangle|^2 + |\langle T_j^* x, y\rangle|^2}{2}\right]$$

$$\leq \frac{1}{2}\sum_{j=1}^{n} p_j \left[\|T_j x\| \|T_j^* y\| + \|T_j y\| \|T_j^* x\|\right]$$

$$\leq \frac{1}{2}\left(\sum_{j=1}^{n} p_j \|T_j x\|^2\right)^{1/2}\left(\sum_{j=1}^{n} p_j \|T_j^* y\|\right)^{1/2}$$

$$+ \frac{1}{2}\left(\sum_{j=1}^{n} p_j \|T_j y\|^2\right)^{1/2}\left(\sum_{j=1}^{n} p_j \|T_j^* x\|^2\right)^{1/2}$$

$$\leq \left(\sum_{j=1}^{n} p_j \left[\frac{\|T_j x\|^2 + \|T_j y\|^2}{2}\right]\right)^{1/2}\left(\sum_{j=1}^{n} p_j \left[\frac{\|T_j^* y\|^2 + \|T_j^* x\|^2}{2}\right]\right)^{1/2}$$

$$\leq \sum_{j=1}^{n} p_j \left[\frac{\left\| T_j x \right\|^2 + \left\| T_j y \right\|^2 + \left\| T_j^* y \right\|^2 + \left\| T_j^* x \right\|^2}{4} \right]$$

$$= \frac{1}{2} \left[\sum_{j=1}^{n} p_j \left\langle \frac{\left| T_j \right|^2 + \left| T_j^* \right|^2}{2} x, x \right\rangle + \sum_{j=1}^{n} p_j \left\langle \frac{\left| T_j \right|^2 + \left| T_j^* \right|^2}{2} y, y \right\rangle \right]$$

for any $x, y \in H$ with $\|x\| = \|y\| = 1$.

Remark 2.38 The case of normal operators N_j, $j \in \{1, ..., n\}$ is of interest for functions of operators and maybe stated as follows:

$$\sum_{j=1}^{n} p_j \left[\left| \left\langle \left(\frac{N_j + N_j^*}{2} \right) x, y \right\rangle \right|^2 \right] \qquad (2.132)$$

$$\leq \sum_{j=1}^{n} p_j \left[\frac{\left| \langle N_j x, y \rangle \right|^2 + \left| \langle N_j^* x, y \rangle \right|^2}{2} \right]$$

$$\leq \frac{1}{2} \sum_{j=1}^{n} p_j \left[\left\| N_j x \right\|^{2\alpha} \left\| N_j y \right\|^{2(1-\alpha)} + \left\| N_j y \right\|^{2\alpha} \left\| N_j x \right\|^{2(1-\alpha)} \right]$$

$$\leq \frac{1}{2} \left(\sum_{j=1}^{n} p_j \left\| N_j x \right\|^2 \right)^{\alpha} \left(\sum_{j=1}^{n} p_j \left\| N_j y \right\|^2 \right)^{1-\alpha}$$

$$+ \frac{1}{2} \left(\sum_{j=1}^{n} p_j \left\| N_j y \right\|^2 \right)^{\alpha} \left(\sum_{j=1}^{n} p_j \left\| N_j x \right\|^2 \right)^{1-\alpha}$$

$$\leq \frac{1}{2} \sum_{j=1}^{n} p_j \left[\left\| N_j x \right\|^2 + \left\| N_j y \right\|^2 \right]$$

for any $x, y \in H$ with $\|x\| = \|y\| = 1$ and $\alpha \in [0, 1]$.

2.10   Other Inequalities for Functions of Normal Operators

Now, by the help of power series $f(z) = \sum_{n=0}^{\infty} a_n z^n$ we can naturally construct another power series which will have as coefficients the absolute values of the coefficient of the original series, namely, $f_A(z) := \sum_{n=0}^{\infty} |a_n| z^n$. It is obvious that this new power series will have the same radius of convergence as the original series. We also notice that if all coefficients $a_n \geq 0$, then $f_A = f$.

The following result is a functional inequality for normal operators that can be obtained from (2.110).

Theorem 2.39 (Dragomir et al. [10]) *Let $f(z) = \sum_{n=0}^{\infty} a_n z^n$ be a function defined by power series with complex coefficients and convergent on the open disk $D(0, R) \subset \mathbb{C}$, $R > 0$. If N is a normal operator on the Hilbert space H and for $\alpha \in [0, 1]$ we have that $\|N\|^{2\alpha}, \|N\|^{2(1-\alpha)} < R$, then we have the inequalities*

$$\left| \left\langle \left(\frac{f(N) + f(N^*)}{2} \right) x, y \right\rangle \right| \leq \left\langle \left(\frac{f_A(|N|^{2\alpha}) + f(|N|^{2(1-\alpha)})}{2} \right) x, x \right\rangle^{1/2} \qquad (2.133)$$

$$\times \left\langle \left(\frac{f_A(|N|^{2\alpha}) + f(|N|^{2(1-\alpha)})}{2} \right) y, y \right\rangle^{1/2}$$

for any $x, y \in H$.

Proof If N is a normal operator, then for any $j \in \mathbb{N}$ we have that

$$\left| N^j \right|^2 = (N^* N)^j = |N|^{2j}.$$

Utilising the inequality (2.120) we have

$$\left| \left\langle \sum_{j=0}^{n} a_j \left(\frac{N^j + (N^*)^j}{2} \right) x, y \right\rangle \right| \qquad (2.134)$$

$$\leq \sum_{j=0}^{n} |a|_j \left| \left\langle \frac{N^j + (N^*)^j}{2} x, y \right\rangle \right|$$

$$\leq \sum_{j=0}^{n} |a|_j \left[\frac{|\langle N^j x, y \rangle| + |\langle (N^*)^j x, y \rangle|}{2} \right]$$

$$\leq \left\langle \sum_{j=0}^{n} |a|_j \left[\frac{(|N|^{2\alpha})^j + (|N|^{2(1-\alpha)})^j}{2} \right] x, x \right\rangle^{1/2}$$

$$\times \left\langle \sum_{j=1}^{n} |a|_0 \left[\frac{(|N|^{2\alpha})^j + (|N|^{2(1-\alpha)})^j}{2} \right] y, y \right\rangle^{1/2}$$

for any $\alpha \in [0, 1]$, $n \in \mathbb{N}$ and any $x, y \in H$.

Since $\|N\|^{2\alpha}, \|N\|^{2(1-\alpha)} < R$, then it follows that the series $\sum_{j=0}^{\infty} |a_j| \left(|N|^{2\alpha} \right)^j$ and $\sum_{j=0}^{\infty} |a_j| \left(|N|^{2(1-\alpha)} \right)^j$ are absolute convergent in $\mathcal{B}(H)$, and by taking the limit over $n \to \infty$ in (2.134) we deduce the desired result (2.133). ∎

Remark 2.40 With the assumptions in Theorem 2.39, if we take the supremum over $y \in H$, $\|y\| = 1$, then we get the vector inequality

$$\left\| \left(\frac{f(N) + f(N^*)}{2} \right) x \right\| \leq \frac{1}{2} \langle (f_A(|N|^{2\alpha}) + f(|N|^{2(1-\alpha)})) x, x \rangle^{1/2} \quad (2.135)$$
$$\times \| f_A(|N|^{2\alpha}) + f(|N|^{2(1-\alpha)}) \|$$

for any $x \in H$, which in its turn produces the norm inequality

$$\left\| \frac{f(N) + f(N^*)}{2} \right\| \leq \frac{1}{2} \| f_A(|N|^{2\alpha}) + f(|N|^{2(1-\alpha)}) \| \quad (2.136)$$

for any $\alpha \in [0, 1]$.

Moreover, if we take $y = x$ in (2.133), then we have

$$\left| \left\langle \frac{f(N) + f(N^*)}{2} x, x \right\rangle \right| \leq \frac{1}{2} \langle [f_A(|N|^{2\alpha}) + f(|N|^{2(1-\alpha)})] x, x \rangle \quad (2.137)$$

for any $x \in H$, which, by taking the supremum over $x \in H$, $\|x\| = 1$ generates the numerical radius inequality

$$w \left(\frac{f(N) + f(N^*)}{2} \right) \leq \frac{1}{2} w [f_A(|N|^{2\alpha}) + f(|N|^{2(1-\alpha)})] \quad (2.138)$$

for any $\alpha \in [0, 1]$.

We can state the particular vector inequalities:

$$\left| \left\langle \left[\frac{\ln(1_H + N)^{-1} + \ln(1_H + N^*)^{-1}}{2} \right] x, y \right\rangle \right| \quad (2.139)$$
$$\leq \frac{1}{2} \left\langle \left[\ln \left(1_H - |N|^{2\alpha} \right)^{-1} + \ln \left(1_H - |N|^{2(1-\alpha)} \right)^{-1} \right] x, x \right\rangle^{1/2}$$
$$\times \left\langle \left[\ln \left(1_H - |N|^{2\alpha} \right)^{-1} + \ln \left(1_H - |N|^{2(1-\alpha)} \right)^{-1} \right] y, y \right\rangle^{1/2},$$

and

$$\left| \left\langle \left[\frac{(1_H + N)^{-1} + (1_H + N^*)^{-1}}{2} \right] x, y \right\rangle \right| \quad (2.140)$$
$$\leq \frac{1}{2} \left\langle \left[\left(1_H - |N|^{2\alpha} \right)^{-1} + \left(1_H - |N|^{2(1-\alpha)} \right)^{-1} \right] x, x \right\rangle^{1/2}$$
$$\times \left\langle \left[\ln \left(1_H - |N|^{2\alpha} \right)^{-1} + \ln \left(1_H - |N|^{2(1-\alpha)} \right)^{-1} \right] y, y \right\rangle^{1/2},$$

for any $x, y \in H$ and $\|N\| < 1$.

We also have the inequalities

$$\left|\left\langle \left[\frac{\sin(N)+\sin(N^*)}{2}\right]x,y\right\rangle\right| \tag{2.141}$$

$$\leq \frac{1}{2}\left\langle \left[\sinh\left(|N|^{2\alpha}\right)+\sinh\left(|N|^{2(1-\alpha)}\right)\right]x,x\right\rangle^{1/2}$$
$$\times \left\langle \left[\sinh\left(|N|^{2\alpha}\right)+\sinh\left(|N|^{2(1-\alpha)}\right)\right]y,y\right\rangle^{1/2}$$

and

$$\left|\left\langle \left[\frac{\cos(N)+\cos(N^*)}{2}\right]x,y\right\rangle\right| \tag{2.142}$$

$$\leq \frac{1}{2}\left\langle \left[\cosh\left(|N|^{2\alpha}\right)+\cosh\left(|N|^{2(1-\alpha)}\right)\right]x,x\right\rangle^{1/2}$$
$$\times \left\langle \left[\cosh\left(|N|^{2\alpha}\right)+\cosh\left(|N|^{2(1-\alpha)}\right)\right]y,y\right\rangle^{1/2}$$

for any $x,y \in H$ and N a normal operator.

If we utilize function as power series representations with nonnegative coefficients, then we can state the following vector inequalities as well:

$$\left|\left\langle \left[\frac{\exp(N)+\exp(N^*)}{2}\right]x,y\right\rangle\right| \tag{2.143}$$

$$\leq \frac{1}{2}\left\langle \left[\exp\left(|N|^{2\alpha}\right)+\exp\left(|N|^{2(1-\alpha)}\right)\right]x,x\right\rangle^{1/2}$$
$$\times \left\langle \left[\exp\left(|N|^{2\alpha}\right)+\exp\left(|N|^{2(1-\alpha)}\right)\right]y,y\right\rangle^{1/2}$$

for any $x,y \in H$ and N a normal operator.

If $\|N\| < 1$, then we also have the inequalities

$$\left|\left\langle \left[\frac{\ln\left(\frac{1_H+N}{1_H-N}\right)+\ln\left(\frac{1_H+N^*}{1_H-N^*}\right)}{2}\right]x,y\right\rangle\right| \tag{2.144}$$

$$\leq \frac{1}{2}\left\langle \left[\ln\left(\frac{1_H+|N|^{2\alpha}}{1_H-|N|^{2\alpha}}\right)+\ln\left(\frac{1_H+|N|^{2(1-\alpha)}}{1_H-|N|^{2(1-\alpha)}}\right)\right]x,x\right\rangle^{1/2}$$
$$\times \left\langle \left[\ln\left(\frac{1_H+|N|^{2\alpha}}{1_H-|N|^{2\alpha}}\right)+\ln\left(\frac{1_H+|N|^{2(1-\alpha)}}{1_H-|N|^{2(1-\alpha)}}\right)\right]y,y\right\rangle^{1/2}$$

$$\left| \left\langle \left[\frac{\tanh^{-1}(N) + \tanh^{-1}(N^*)}{2} \right] x, y \right\rangle \right| \tag{2.145}$$

$$\leq \frac{1}{2} \left\langle \left[\tanh^{-1}\left(|N|^{2\alpha}\right) + \tanh^{-1}\left(|N|^{2(1-\alpha)}\right) \right] x, x \right\rangle^{1/2}$$

$$\times \left\langle \left[\tanh^{-1}\left(|N|^{2\alpha}\right) + \tanh^{-1}\left(|N|^{2(1-\alpha)}\right) \right] y, y \right\rangle^{1/2}$$

and

$$\left| \left\langle \left[\frac{{}_2F_1(\alpha, \beta, \gamma, N) + {}_2F_1(\alpha, \beta, \gamma, N^*)}{2} \right] x, y \right\rangle \right| \tag{2.146}$$

$$\leq \frac{1}{2} \left\langle \left[{}_2F_1\left(\alpha, \beta, \gamma, |N|^{2\alpha}\right) + {}_2F_1\left(\alpha, \beta, \gamma, |N|^{2(1-\alpha)}\right) \right] x, x \right\rangle^{1/2}$$

$$\times \left\langle \left[{}_2F_1\left(\alpha, \beta, \gamma, |N|^{2\alpha}\right) + {}_2F_1\left(\alpha, \beta, \gamma, |N|^{2(1-\alpha)}\right) \right] y, y \right\rangle^{1/2}$$

for any $x, y \in H$.

From a different perspective, we also have:

Theorem 2.41 (Dragomir et al. [10]) *With the assumption of Theorem 2.39 and if N is a normal operator on the Hilbert space H and $z \in \mathbb{C}$ such that $\|N\|^2$, $|z|^2 < R$, then we have the inequalities*

$$\left| \left\langle \left(\frac{f(zN) + f(zN^*)}{2} \right) x, y \right\rangle \right|^2 \tag{2.147}$$

$$\leq \frac{1}{2} f_A\left(|z|^2\right) \left[\langle f_A\left(|N|^2\right) x, x \rangle^\alpha \langle f_A\left(|N|^2\right) y, y \rangle^{1-\alpha} \right.$$

$$\left. + \langle f_A\left(|N|^2\right) y, y \rangle^\alpha \langle f_A\left(|N|^2\right) x, x \rangle^{1-\alpha} \right]$$

$$\leq \frac{1}{2} f_A\left(|z|^2\right) \left[\langle f_A\left(|N|^2\right) x, x \rangle + \langle f_A\left(|N|^2\right) y, y \rangle \right]$$

for any $x, y \in H$ with $\|x\| = \|y\| = 1$ and $\alpha \in [0, 1]$.

In particular, for $\alpha = \frac{1}{2}$ we have

$$\left| \left\langle \left(\frac{f(zN) + f(zN^*)}{2} \right) x, y \right\rangle \right|^2 \tag{2.148}$$

$$\leq f_A\left(|z|^2\right) \langle f_A\left(|N|^2\right) x, x \rangle^{1/2} \langle f_A\left(|N|^2\right) y, y \rangle^{1/2}$$

$$\leq \frac{1}{2} f_A\left(|z|^2\right) \left[\langle f_A\left(|N|^2\right) x, x \rangle + \langle f_A\left(|N|^2\right) y, y \rangle \right]$$

for any $x, y \in H$ with $\|x\| = \|y\| = 1$.

Proof If we use the third and the fourth inequality in (2.132) we have

$$\sum_{j=0}^{n} |a_j| \left[\left| \left\langle \left(\frac{N^j + (N^*)^j}{2} \right) x, y \right\rangle \right|^2 \right] \tag{2.149}$$

$$\leq \frac{1}{2} \left(\sum_{j=0}^{n} |a_j| \, \|N^j x\|^2 \right)^{\alpha} \left(\sum_{j=0}^{n} |a_j| \, \|N^j y\|^2 \right)^{1-\alpha}$$

$$+ \frac{1}{2} \left(\sum_{j=0}^{n} |a_j| \, \|N^j y\|^2 \right)^{\alpha} \left(\sum_{j=0}^{n} |a_j| \, \|N^j x\|^2 \right)^{1-\alpha}$$

$$\leq \frac{1}{2} \sum_{j=0}^{n} |a_j| \left[\|N^j x\|^2 + \|N^j y\|^2 \right]$$

for any $x, y \in H$ with $\|x\| = \|y\| = 1$ and $\alpha \in [0, 1]$.

Since N is a normal operator on the Hilbert space H, then

$$\|N^j x\|^2 = \left\langle |N^j|^2 x, x \right\rangle = \left\langle |N|^{2j} x, x \right\rangle$$

for any $j \in \{0, ..., n\}$ and for any $x \in H$ with $\|x\| = 1$.

Then from (2.149) we get

$$\sum_{j=0}^{n} |a_j| \left[\left| \left\langle \left(\frac{N^j + (N^*)^j}{2} \right) x, y \right\rangle \right|^2 \right] \tag{2.150}$$

$$\leq \frac{1}{2} \left(\left\langle \sum_{j=0}^{n} |a_j| \, |N|^{2j} x, x \right\rangle \right)^{\alpha} \left(\left\langle \sum_{j=0}^{n} |a_j| \, |N|^{2j} y, y \right\rangle \right)^{1-\alpha}$$

$$+ \frac{1}{2} \left(\left\langle \sum_{j=0}^{n} |a_j| \, |N|^{2j} y, y \right\rangle \right)^{\alpha} \left(\left\langle \sum_{j=0}^{n} |a_j| \, |N|^{2j} x, x \right\rangle \right)^{1-\alpha}$$

$$\leq \frac{1}{2} \left[\left\langle \sum_{j=0}^{n} |a_j| \, |N|^{2j} x, x \right\rangle + \left\langle \sum_{j=0}^{n} |a_j| \, |N|^{2j} y, y \right\rangle \right]$$

for any $x, y \in H$ with $\|x\| = \|y\| = 1$ and $\alpha \in [0, 1]$.

By the weighted Cauchy–Buniakowski–Schwarz inequality we also have

$$
\left| \left\langle \sum_{j=0}^{n} a_j z^j \left(\frac{N^j + (N^*)^j}{2} \right) x, y \right\rangle \right|^2 \tag{2.151}
$$

$$
\leq \sum_{j=0}^{n} |a_j| |z|^{2j} \sum_{j=0}^{n} |a_j| \left[\left| \left\langle \left(\frac{N^j + (N^*)^j}{2} \right) x, y \right\rangle \right|^2 \right]
$$

for any $x, y \in H$ with $\|x\| = \|y\| = 1$.

Now, since the series $\sum_{j=0}^{\infty} a_j z^j N^j$, $\sum_{j=0}^{\infty} a_j z^j (N^*)^j$, $\sum_{j=0}^{\infty} |a_j| |z|^{2j}$, $\sum_{j=0}^{\infty} |a_j| |N|^{2j}$ are convergent, then by (2.150) and (2.151) on letting $n \to \infty$, we deduce the desired result (2.147). ∎

Similar inequalities for some particular functions of interest can be stated. However, the details are left to the interested reader.

2.11 Examples for the Euclidian Norm

We have:

Theorem 2.42 (Dragomir et al. [10]) *For any* $(T_1, \ldots, T_n) \in B^{(n)}(H)$ *we have*

$$
\left\| \left(\frac{T_1 + T_1^*}{2}, \ldots, \frac{T_n + T_n^*}{2} \right) \right\|_e^2 \leq \left\| \sum_{j=1}^{n} |T_j|^2 \right\|^{\alpha} \left\| \sum_{j=1}^{n} |T_j^*|^2 \right\|^{1-\alpha} \tag{2.152}
$$

$$
\leq \alpha \left\| \sum_{j=1}^{n} |T_j|^2 \right\| + (1 - \alpha) \left\| \sum_{j=1}^{n} |T_j^*|^2 \right\|
$$

and

$$
w_e^2 \left(\frac{T_1 + T_1^*}{2}, \ldots, \frac{T_n + T_n^*}{2} \right) \tag{2.153}
$$

$$
\leq \sup_{\|x\|=1} \left[\left\langle \sum_{j=1}^{n} |T_j|^2 x, x \right\rangle^{\alpha} \left\langle \sum_{j=1}^{n} |T_j^*|^2 x, x \right\rangle^{1-\alpha} \right]
$$

$$\leq \begin{cases} \left\| \sum_{j=1}^{n} |T_j|^2 \right\|^{\alpha} \left\| \sum_{j=1}^{n} |T_j^*|^2 \right\|^{1-\alpha} \\[2em] \left\| \alpha \sum_{j=1}^{n} |T_j|^2 + (1-\alpha) \sum_{j=1}^{n} |T_j^*|^2 \right\| \end{cases}$$

$$\leq \alpha \left\| \sum_{j=1}^{n} |T_j|^2 \right\| + (1-\alpha) \left\| \sum_{j=1}^{n} |T_j^*|^2 \right\|$$

for any $\alpha \in [0, 1]$.

Proof Making use of the inequalities (2.122) and (2.129) we have

$$\sum_{j=1}^{n} \left[\left| \left\langle \left(\frac{T_j + T_j^*}{2} \right) x, y \right\rangle \right|^2 \right] \qquad (2.154)$$

$$\leq \frac{1}{2} \left\langle \sum_{j=1}^{n} |T_j|^2 x, x \right\rangle^{\alpha} \left\langle \sum_{j=1}^{n} |T_j^*|^2 y, y \right\rangle^{1-\alpha}$$

$$+ \frac{1}{2} \left\langle \sum_{j=1}^{n} |T_j|^2 y, y \right\rangle^{\alpha} \left\langle \sum_{j=1}^{n} |T_j^*|^2 x, x \right\rangle^{1-\alpha}$$

for any $x, y \in H$ with $\|x\| = \|y\| = 1$ and $\alpha \in [0, 1]$.

Taking the supremum over $\|x\| = \|y\| = 1$ in (2.154) we get

$$\left\| \left(\frac{T_1 + T_1^*}{2}, \dots, \frac{T_n + T_n^*}{2} \right) \right\|_e^2$$

$$\leq \frac{1}{2} \sup_{\|x\|=1} \left\langle \sum_{j=1}^{n} |T_j|^2 x, x \right\rangle^{\alpha} \sup_{\|y\|=1} \left\langle \sum_{j=1}^{n} |T_j^*|^2 y, y \right\rangle^{1-\alpha}$$

$$+ \frac{1}{2} \sup_{\|y\|=1} \left\langle \sum_{j=1}^{n} |T_j|^2 y, y \right\rangle^{\alpha} \sup_{\|x\|=1} \left\langle \sum_{j=1}^{n} |T_j^*|^2 x, x \right\rangle^{1-\alpha}$$

$$= \left\| \sum_{j=1}^{n} |T_j|^2 \right\|^{\alpha} \left\| \sum_{j=1}^{n} |T_j^*|^2 \right\|^{1-\alpha}$$

and the inequality (2.152) is proved.

Now, if we take $y = x$ in (2.154) we get

$$\sum_{j=1}^{n} \left[\left| \left\langle \left(\frac{T_j + T_j^*}{2} \right) x, x \right\rangle \right|^2 \right] \tag{2.155}$$

$$\leq \left\langle \sum_{j=1}^{n} |T_j|^2 x, x \right\rangle^{\alpha} \left\langle \sum_{j=1}^{n} |T_j^*|^2 x, x \right\rangle^{1-\alpha}$$

$$\leq \left\langle \left[\alpha \sum_{j=1}^{n} |T_j|^2 + (1-\alpha) \sum_{j=1}^{n} |T_j^*|^2 \right] x, x \right\rangle$$

for any $x \in H$ with $\|x\| = 1$ and $\alpha \in [0, 1]$.

Taking the supremum over $\|x\| = 1$ in (2.155) we get the desired result. ∎

Remark 2.43 In the particular case $\alpha = \frac{1}{2}$ we get

$$\left\| \left(\frac{T_1 + T_1^*}{2}, \ldots, \frac{T_n + T_n^*}{2} \right) \right\|_e^2 \leq \left\| \sum_{j=1}^{n} |T_j|^2 \right\|^{1/2} \left\| \sum_{j=1}^{n} |T_j^*|^2 \right\|^{1/2} \tag{2.156}$$

$$\leq \frac{1}{2} \left[\left\| \sum_{j=1}^{n} |T_j|^2 \right\| + \left\| \sum_{j=1}^{n} |T_j^*|^2 \right\| \right]$$

and

$$w_e^2 \left(\frac{T_1 + T_1^*}{2}, \ldots, \frac{T_n + T_n^*}{2} \right) \tag{2.157}$$

$$\leq \sup_{\|x\|=1} \left[\left\langle \sum_{j=1}^{n} |T_j|^2 x, x \right\rangle^{1/2} \left\langle \sum_{j=1}^{n} |T_j^*|^2 x, x \right\rangle^{1/2} \right]$$

$$\leq \begin{cases} \left\| \sum_{j=1}^{n} |T_j|^2 \right\|^{1/2} \left\| \sum_{j=1}^{n} |T_j^*|^2 \right\|^{1/2} \\ \\ \left\| \sum_{j=1}^{n} \frac{|T_j|^2 + |T_j^*|^2}{2} \right\| \end{cases}$$

$$\leq \frac{1}{2} \left[\left\| \sum_{j=1}^{n} |T_j|^2 \right\| + \left\| \sum_{j=1}^{n} |T_j^*|^2 \right\| \right].$$

2.12 Examples for s-1-Norm and s-1-Numerical Radius

We have:

Theorem 2.44 (Dragomir et al. [10]) *For any* $(T_1, \ldots, T_n) \in B^{(n)}(H)$ *we have*

$$\left\| \left(\frac{T_1 + T_1^*}{2}, \ldots, \frac{T_n + T_n^*}{2} \right) \right\|_{s,1} \leq \left\| \sum_{j=1}^{n} \left[\frac{|T_j|^{2\alpha} + |T_j^*|^{2(1-\alpha)}}{2} \right] \right\| \tag{2.158}$$

$$\leq \frac{1}{2} \left[\left\| \sum_{j=1}^{n} |T_j|^{2\alpha} \right\| + \left\| \sum_{j=1}^{n} |T_j^*|^{2(1-\alpha)} \right\| \right]$$

and

$$w_{s,1} \left(\frac{T_1 + T_1^*}{2}, \ldots, \frac{T_n + T_n^*}{2} \right) \leq w_{s,1} (T_1, \ldots, T_n) \tag{2.159}$$

$$\leq \left\| \sum_{j=1}^{n} \left[\frac{|T_j|^{2\alpha} + |T_j^*|^{2(1-\alpha)}}{\cdot \, 2} \right] \right\|$$

$$\leq \frac{1}{2} \left[\left\| \sum_{j=1}^{n} |T_j|^{2\alpha} \right\| + \left\| \sum_{j=1}^{n} |T_j^*|^{2(1-\alpha)} \right\| \right]$$

for any $\alpha \in [0, 1]$.

Proof Utilizing the inequality (2.110) we have

$$\sum_{j=1}^{n} \left| \left\langle \frac{T_j + T_j^*}{2} x, y \right\rangle \right| \tag{2.160}$$

$$\leq \left\langle \sum_{j=1}^{n} \left[\frac{|T_j|^{2\alpha} + |T_j^*|^{2(1-\alpha)}}{2} \right] x, x \right\rangle^{1/2}$$

$$\times \left\langle \sum_{j=1}^{n} \left[\frac{|T_j|^{2\alpha} + |T_j^*|^{2(1-\alpha)}}{2} \right] y, y \right\rangle^{1/2}$$

for any $x, y \in H$ and $\alpha \in [0, 1]$.

Taking the supremum in (2.160) over $\|x\| = \|y\| = 1$ we get the first inequality in (2.158).

The second part follows by the triangle inequality.

By the inequality (2.116) we have

$$\sum_{j=1}^{n} \left| \left\langle \frac{T_j + T_j^*}{2} x, x \right\rangle \right| \le \sum_{j=1}^{n} |\langle T_j x, x \rangle|$$

$$\le \left\langle \sum_{j=1}^{n} \left[\frac{|T_j|^{2\alpha} + \left|T_j^*\right|^{2(1-\alpha)}}{2} \right] x, x \right\rangle$$

for any $x \in H$.

Taking the supremum over $\|x\| = 1$ we deduce the desired result (2.159). ∎

Remark 2.45 The case $\alpha = \frac{1}{2}$ produces the following chains of inequalities

$$\left\| \sum_{j=1}^{n} \left(\frac{T_j + T_j^*}{2} \right) \right\| \le \left\| \left(\frac{T_1 + T_1^*}{2}, \ldots, \frac{T_n + T_n^*}{2} \right) \right\|_{s,1} \qquad (2.161)$$

$$\le \left\| \sum_{j=1}^{n} \left(\frac{|T_j| + \left|T_j^*\right|}{2} \right) \right\|$$

$$\le \frac{1}{2} \left[\left\| \sum_{j=1}^{n} |T_j| \right\| + \left\| \sum_{j=1}^{n} |T_j^*| \right\| \right]$$

and

$$w \left(\sum_{j=1}^{n} \left(\frac{T_j + T_j^*}{2} \right) \right) \le w_{s,1} \left(\frac{T_1 + T_1^*}{2}, \ldots, \frac{T_n + T_n^*}{2} \right) \qquad (2.162)$$

$$\le w_{s,1} (T_1, \ldots, T_n)$$

$$\le \left\| \sum_{j=1}^{n} \left(\frac{|T_j| + \left|T_j^*\right|}{2} \right) \right\|$$

$$\le \frac{1}{2} \left[\left\| \sum_{j=1}^{n} |T_j| \right\| + \left\| \sum_{j=1}^{n} |T_j^*| \right\| \right].$$

Chapter 3
Generalizations of Furuta's Type

In this chapter we present a two parameter generalization of Kato due to Furuta. Applications for functions of bounded linear operators defined by power series and inequalities for four bounded operators generalizing Furuta's inequality and provide some general *Norm* and *Numerical Radius* inequalities are given as well.

3.1 Furuta's Inequality

In the following we denote by $\mathcal{B}(H)$ the *Banach algebra* of all bounded linear operators on a complex Hilbert space $(H; \langle \cdot, \cdot \rangle)$.

In 1952, Kato [22] proved the following celebrated generalization of Schwarz inequality for any bounded linear operator T on H:

$$|\langle Tx, y \rangle|^2 \le \langle (T^*T)^\alpha x, x \rangle \langle (TT^*)^{1-\alpha} y, y \rangle, \tag{K}$$

for any $x, y \in H$ and $\alpha \in [0, 1]$. Utilizing the modulus notation, we can write (K) as follows

$$|\langle Tx, y \rangle|^2 \le \langle |T|^{2\alpha} x, x \rangle \langle |T^*|^{2(1-\alpha)} y, y \rangle \tag{3.1}$$

for any $x, y \in H$ and $\alpha \in [0, 1]$.

In order to generalize this result, in 1994 Furuta [19] obtained the following result:

$$\left|\langle T |T|^{\alpha+\beta-1} x, y \rangle\right|^2 \le \langle |T|^{2\alpha} x, x \rangle \langle |T^*|^{2\beta} y, y \rangle \tag{F}$$

for any $x, y \in H$ and $\alpha, \beta \in [0, 1]$ with $\alpha + \beta \ge 1$.

© The Author(s), under exclusive license to Springer Nature Switzerland AG 2019
S. S. Dragomir, *Kato's Type Inequalities for Bounded Linear
Operators in Hilbert Spaces*, SpringerBriefs in Mathematics,
https://doi.org/10.1007/978-3-030-17459-0_3

From the proof in [19], one realizes that the condition $\alpha, \beta \in [0, 1]$ is taken only to fit with the result from the *Heinz-Kato inequality*

$$|\langle Tx, y\rangle| \leq \|A^\alpha x\| \, \|B^{1-\alpha} y\|, \tag{HK}$$

for any $x, y \in H$ and $\alpha \in [0, 1]$, where A and B are positive operators such that $\|Tx\| \leq \|Ax\|$ and $\|T^*y\| \leq \|By\|$ for all $x, y \in H$.

Therefore, one can state the more general result:

Theorem 3.1 (Furuta Inequality [19]) *Let* $T \in \mathcal{B}(H)$ *and* $\alpha, \beta \geq 0$ *with* $\alpha + \beta \geq 1$. *Then for any* $x, y \in H$ *we have the inequality* (F).

We observe that this fact allows for some particular instances of interest that were not possible in the case when $\alpha, \beta \in [0, 1]$.

If we take $\beta = \alpha$ in (F) then we get

$$\left|\langle T \, |T|^{2\alpha-1} x, y\rangle\right|^2 \leq \langle |T|^{2\alpha} x, x\rangle \left\langle |T^*|^{2\alpha} y, y\right\rangle \tag{3.2}$$

for any $x, y \in H$ and $\alpha \geq \frac{1}{2}$. In particular, for $\alpha = 1$ we get

$$|\langle T \, |T| x, y\rangle|^2 \leq \langle |T|^2 x, x\rangle \left\langle |T^*|^2 y, y\right\rangle \tag{3.3}$$

for any $x, y \in H$.

If we take $T = N$ a *normal operator*, i.e., we recall that $NN^* = N^*N$, then we get from (F) the following inequality for normal operators

$$\left|\langle N \, |N|^{\alpha+\beta-1} x, y\rangle\right|^2 \leq \langle |N|^{2\alpha} x, x\rangle \langle |N|^{2\beta} y, y\rangle \tag{3.4}$$

for any $x, y \in H$ and $\alpha, \beta \geq 0$ with $\alpha + \beta \geq 1$.

This implies the inequalities

$$\left|\langle N \, |N|^{2\alpha-1} x, y\rangle\right|^2 \leq \langle |N|^{2\alpha} x, x\rangle \langle |N|^{2\alpha} y, y\rangle \tag{3.5}$$

for any $x, y \in H$ and $\alpha \geq \frac{1}{2}$ and, in particular,

$$|\langle N \, |N| x, y\rangle|^2 \leq \langle |N|^2 x, x\rangle \langle |N|^2 y, y\rangle \tag{3.6}$$

for any $x, y \in H$.

Making $y = x$ in (3.5) produces

$$\left|\langle N \, |N|^{2\alpha-1} x, x\rangle\right| \leq \langle |N|^{2\alpha} x, x\rangle$$

for any $x \in H$ and $\alpha \geq \frac{1}{2}$ and, in particular,

$$|\langle N |N| x, x\rangle| \le \langle |N|^2 x, x\rangle$$

for any $x \in H$.

3.2 Functional Generalizations

Now, by the help of power series $f(z) = \sum_{n=0}^{\infty} a_n z^n$ we can naturally construct another power series which has as coefficients the absolute values of the coefficients of the original series, namely, $f_A(z) := \sum_{n=0}^{\infty} |a_n| z^n$. It is obvious that this new power series has the same radius of convergence as the original series. We also notice that if all coefficients $a_n \ge 0$, then $f_A = f$.

Theorem 3.2 (Dragomir [4]) *Let* $f(z) = \sum_{n=0}^{\infty} a_n z^n$ *be a function defined by power series with real coefficients and convergent on the open disk* $D(0, R) \subset \mathbb{C}$, $R > 0$. *Let* $T \in \mathcal{B}(H)$, $\alpha, \beta \ge 0$ *with* $\alpha + \beta \ge 1$ *and such that*

$$\|T\|^{2\alpha}, \|T\|^{2\beta} < R, \qquad (3.7)$$

then we have the inequality

$$\left|\langle Tf\left(|T|^{\alpha+\beta}\right) |T|^{\alpha+\beta-1} x, y\rangle\right|^2 \qquad (3.8)$$
$$\le \langle f_A\left(|T|^{2\alpha}\right) |T|^{2\alpha} x, x\rangle \langle f_A\left(|T^*|^{2\beta}\right) |T^*|^{2\beta} y, y\rangle$$

for any $x, y \in H$.

Proof Since $\alpha, \beta \ge 0$ with $\alpha + \beta \ge 1$, then $n\alpha + n\beta \ge 1$ for any $n \ge 1$.

From Furuta's inequality (F) we have the power inequality

$$\left|\langle T |T|^{n\alpha+n\beta-1} x, y\rangle\right| \le \langle |T|^{2n\alpha} x, x\rangle^{1/2} \langle |T^*|^{2n\beta} y, y\rangle^{1/2}, \qquad (3.9)$$

for any natural numbers $n \ge 1$ and $x, y \in H$.

If we multiply this inequality with the positive quantities $|a_{n-1}|$, use the triangle inequality and the Cauchy–Bunyakowsky–Schwarz discrete inequality we have successively

$$\left|\langle \sum_{n=1}^{k} a_{n-1} T |T|^{n(\alpha+\beta)-1} x, y\rangle\right| \qquad (3.10)$$
$$\le \sum_{n=1}^{k} |a_{n-1}| \left|\langle T |T|^{n(\alpha+\beta)-1} x, y\rangle\right|$$

$$\leq \sum_{n=1}^{k} |a_{n-1}| \left\langle |T|^{2n\alpha} x, x \right\rangle^{1/2} \left\langle |T^*|^{2n\beta} y, y \right\rangle^{1/2}$$

$$\leq \left\langle \sum_{n=1}^{k} |a_{n-1}| |T|^{2n\alpha} x, x \right\rangle^{1/2} \left\langle \sum_{n=1}^{k} |a_{n-1}| |T^*|^{2n\beta} y, y \right\rangle^{1/2}$$

for any $x, y \in H$ and $k \geq 1$.

Observe also that

$$\sum_{n=1}^{k} a_{n-1} T |T|^{n(\alpha+\beta)-1} = T \left(\sum_{n=1}^{k} a_{n-1} |T|^{(n-1)(\alpha+\beta)} \right) |T|^{(\alpha+\beta)-1},$$

$$\sum_{n=1}^{k} |a_{n-1}| |T|^{2n\alpha} = \left(\sum_{n=1}^{k} |a_{n-1}| |T|^{2(n-1)\alpha} \right) |T|^{2\alpha}$$

and

$$\sum_{n=1}^{k} |a_{n-1}| |T^*|^{2n\beta} = \left(\sum_{n=1}^{k} |a_{n-1}| |T^*|^{2(n-1)\beta} \right) |T^*|^{2\beta}$$

for any $k \geq 1$.

Therefore, by (3.10) we have the inequality

$$\left| \left\langle T \left(\sum_{n=1}^{k} a_{n-1} |T|^{(n-1)(\alpha+\beta)} \right) |T|^{(\alpha+\beta)-1} x, y \right\rangle \right|^2 \qquad (3.11)$$

$$\leq \left\langle \left(\sum_{n=1}^{k} |a_{n-1}| |T|^{2(n-1)\alpha} \right) |T|^{2\alpha} x, x \right\rangle$$

$$\times \left\langle \left(\sum_{n=1}^{k} |a_{n-1}| |T^*|^{2(n-1)\beta} \right) |T^*|^{2\beta} y, y \right\rangle$$

for any $x, y \in H$ and $k \geq 1$.

From (3.7) we have that the series $\sum_{n=0}^{\infty} a_n \left(|T|^{\alpha+\beta} \right)^n$, $\sum_{n=0}^{\infty} |a_n| \left(|T|^{2\alpha} \right)^n$ and $\sum_{n=0}^{\infty} |a_n| \left(|T^*|^{2\beta} \right)^n$ are convergent in $\mathcal{B}(H)$ and taking the limit over $k \to \infty$ in (3.11) we deduce the desired result in (3.8). ∎

Corollary 3.3 (Dragomir [4]) *With the assumptions of Theorem 3.2 we have the norm inequality*

$$\left\| Tf \left(|T|^{\alpha+\beta} \right) |T|^{\alpha+\beta-1} \right\|^2 \leq \left\| f_A \left(|T|^{2\alpha} \right) |T|^{2\alpha} \right\| \left\| f_A \left(|T^*|^{2\beta} \right) |T^*|^{2\beta} \right\| \qquad (3.12)$$

and the numerical radius inequality

$$w \left(Tf \left(|T|^{\alpha+\beta} \right) |T|^{\alpha+\beta-1} \right) \leq \frac{1}{2} \left\| f_A \left(|T|^{2\alpha} \right) |T|^{2\alpha} + f_A \left(|T^*|^{2\beta} \right) |T^*|^{2\beta} \right\|.$$

$$(3.13)$$

Proof The proof of (3.12) follows by (3.8) on taking the supremum over $x, y \in H$ with $\|x\| = \|y\| = 1$.

By the inequality (3.8) we also have

$$\left| \left\langle Tf \left(|T|^{\alpha+\beta} \right) |T|^{\alpha+\beta-1} x, x \right\rangle \right|$$

$$\leq \left\langle f_A \left(|T|^{2\alpha} \right) |T|^{2\alpha} x, x \right\rangle^{1/2} \left\langle f_A \left(|T^*|^{2\beta} \right) |T^*|^{2\beta} x, x \right\rangle^{1/2}$$

$$\leq \frac{1}{2} \left\langle \left[f_A \left(|T|^{2\alpha} \right) |T|^{2\alpha} + f_A \left(|T^*|^{2\beta} \right) |T^*|^{2\beta} \right] x, x \right\rangle$$

for any $x \in H$.

Taking the supremum over $\|x\| = 1$ we deduce the desired inequality (3.13). ∎

Remark 3.4 If we take $f(z) = 1$, then we get from (3.8) the Furuta's inequality (F). If we take $\beta = \alpha$ in (3.8), then we get

$$\left| \left\langle Tf \left(|T|^{2\alpha} \right) |T|^{2\alpha-1} x, y \right\rangle \right|^2 \qquad (3.14)$$

$$\leq \left\langle f_A \left(|T|^{2\alpha} \right) |T|^{2\alpha} x, x \right\rangle \left\langle f_A \left(|T^*|^{2\alpha} \right) |T^*|^{2\alpha} y, y \right\rangle$$

provided $\alpha \geq \frac{1}{2}$ and $\|T\|^{2\alpha} < R$.

In particular, we have

$$\left| \left\langle Tf \left(|T| \right) x, y \right\rangle \right|^2 \leq \left\langle f_A \left(|T| \right) |T| x, x \right\rangle \left\langle f_A \left(|T^*| \right) |T^*| y, y \right\rangle \qquad (3.15)$$

for any $T \in B(H)$ with $\|T\| < R$.

Remark 3.5 If we take $\beta = 1 - \alpha$ with $\alpha \in [0, 1]$ in (3.8) then we get the following generalization of Kato's inequality (3.1)

$$\left| \left\langle Tf \left(|T| \right) x, y \right\rangle \right|^2 \qquad (3.16)$$

$$\leq \left\langle f_A \left(|T|^{2\alpha} \right) |T|^{2\alpha} x, x \right\rangle \left\langle f_A \left(|T^*|^{2(1-\alpha)} \right) |T^*|^{2(1-\alpha)} y, y \right\rangle$$

for any $x, y \in H$ and $T \in B(H)$ with $\|T\|^{2\alpha}, \|T\|^{2(1-\alpha)} < R$.

The following result concerning two functions also holds:

Theorem 3.6 (Dragomir [4]) *Let* $f(z) = \sum_{n=0}^{\infty} a_n z^n$ *and be* $g(z) = \sum_{n=0}^{\infty} b_n z^n$ *be two functions defined by power series with real coefficients and both of them convergent on the open disk* $D(0, R) \subset \mathbb{C}$, $R > 0$. *Let* $T \in B(H)$, $\alpha, \beta \geq 0$ *with* $\alpha + \beta \geq 1$ *and* $z, u \in \mathbb{C}$ *such that*

$$|z|^2, |u|^2, \|T\|^{2\alpha}, \|T\|^{2\beta} < R, \tag{3.17}$$

then we have the inequality

$$\left|\langle Tf\left(z\,|T|^{\alpha}\right)g\left(u\,|T|^{\beta}\right)|T|^{\alpha+\beta-1}\,x, y\rangle\right|^2 \tag{3.18}$$

$$\leq f_A\left(|z|^2\right)g_A\left(|u|^2\right)\langle f_A\left(|T|^{2\alpha}\right)|T|^{2\alpha}\,x, x\rangle\langle g_A\left(|T^*|^{2\beta}\right)|T^*|^{2\beta}\,y, y\rangle$$

for any $x, y \in H$.

Proof Since $\alpha, \beta \geq 0$ with $\alpha + \beta \geq 1$, then for any $n, m \geq 1$ natural numbers we also have that $n\alpha + m\beta \geq 1$.

From Furuta's inequality (F) written for $n\alpha + m\beta \geq 1$ we have for any natural numbers $n \geq 1$ and $m \geq 1$ the following power inequality

$$\left|\langle T\,|T|^{n\alpha+m\beta-1}\,x, y\rangle\right| \leq \langle|T|^{2n\alpha}\,x, x\rangle^{1/2}\left(|T^*|^{2m\beta}\,y, y\right)^{1/2}, \tag{3.19}$$

where $x, y \in H$.

If we multiply this inequality with the positive quantities $|a_{n-1}|\,|z|^{n-1}$ and $|b_{m-1}|\,|u|^{m-1}$, use the triangle inequality and the Cauchy–Bunyakowsky–Schwarz discrete inequality we have successively:

$$\left|\sum_{n=1}^{k}\sum_{m=1}^{l} a_{n-1}z^{n-1}b_{m-1}u^{m-1}\langle T\,|T|^{n\alpha+m\beta-1}\,x, y\rangle\right| \tag{3.20}$$

$$\leq \sum_{n=1}^{k}\sum_{m=1}^{l}|a_{n-1}|\,|z|^{n-1}\,|b_{m-1}|\,|u|^{m-1}\left|\langle T\,|T|^{n\alpha+m\beta-1}\,x, y\rangle\right|$$

$$\leq \sum_{n=1}^{k}|a_{n-1}|\,|z|^{n-1}\langle|T|^{2n\alpha}\,x, x\rangle^{1/2}\sum_{m=1}^{l}|b_{m-1}|\,|u|^{m-1}\langle|T^*|^{2m\beta}\,y, y\rangle^{1/2}$$

$$\leq \left(\sum_{n=1}^{k}|a_{n-1}|\,|z|^{2(n-1)}\right)^{1/2}\left\langle\sum_{n=1}^{k}|a_{n-1}|\,|T|^{2n\alpha}\,x, x\right\rangle^{1/2}$$

$$\times \left(\sum_{m=1}^{l}|b_{m-1}|\,|u|^{2(m-1)}\right)^{1/2}\left\langle\sum_{m=1}^{l}|b_{m-1}|\,|T^*|^{2m\beta}\,y, y\right\rangle^{1/2}$$

for any $x, y \in H$ and $k \geq 1, l \geq 1$.

Observe also that

$$\sum_{n=1}^{k}\sum_{m=1}^{l} a_{n-1} z^{n-1} b_{m-1} u^{m-1} \left\langle T \, |T|^{n\alpha+m\beta-1} \, x, y \right\rangle \qquad (3.21)$$

$$= \left\langle T \left(\sum_{n=1}^{k} a_{n-1} z^{n-1} |T|^{(n-1)\alpha} \right) \left(\sum_{m=1}^{l} b_{m-1} u^{m-1} |T|^{(m-1)\beta} \right) |T|^{\alpha+\beta-1} \, x, y \right\rangle$$

for any $x, y \in H$ and $k \geq 1, l \geq 1$.

Making use of (3.20) and (3.21) we get

$$\left| \left\langle T \left(\sum_{n=1}^{k} a_{n-1} z^{n-1} |T|^{(n-1)\alpha} \right) \left(\sum_{m=1}^{l} b_{m-1} u^{m-1} |T|^{(m-1)\beta} \right) |T|^{\alpha+\beta-1} \, x, y \right\rangle \right|$$

$$(3.22)$$

$$\leq \left(\sum_{n=1}^{k} |a_{n-1}| \, |z|^{2(n-1)} \right)^{1/2} \left\langle \left(\sum_{n=1}^{k} |a_{n-1}| \, |T|^{2(n-1)\alpha} \right) |T|^{2\alpha} \, x, x \right\rangle^{1/2}$$

$$\times \left(\sum_{m=1}^{l} |b_{m-1}| \, |u|^{2(m-1)} \right)^{1/2} \left\langle \left(\sum_{m=1}^{l} |b_{m-1}| \, |T^*|^{2(m-1)\beta} \right) |T^*|^{2\beta} \, y, y \right\rangle^{1/2}$$

for any $x, y \in H$ and $k \geq 1, l \geq 1$.

From (3.17) we have that the series

$$\sum_{n=0}^{\infty} a_n z^n \, |T|^{n\alpha} \, , \, \sum_{m=0}^{\infty} b_m u^m \, |T|^{m\beta} \, , \, \sum_{n=0}^{\infty} |a_n| \, |T|^{2n\alpha}$$

and

$$\sum_{m=0}^{\infty} |b_m| \, |T^*|^{2m\beta}$$

are convergent in $\mathcal{B}(H)$ and the series $\sum_{n=0}^{\infty} |a_n| \, |z|^{2n}$ and $\sum_{m=0}^{\infty} |b_m| \, |u|^{2m}$ are convergent in $\mathbb{R}$ and then, by taking the limit over $k \to \infty$ and $l \to \infty$ in (3.22) we deduce desired result (3.18). $\blacksquare$

Remark 3.7 The above inequality (3.18) can provide various particular instances of interest.

For instance, if we take $g = f$ and $z = u$ in Theorem 3.6 then we get

$$\left| \left\langle Tf \left(z \, |T|^{\alpha} \right) f \left(z \, |T|^{\beta} \right) |T|^{\alpha+\beta-1} \, x, y \right\rangle \right|^2 \qquad (3.23)$$

$$\leq f_A^2 \left(|z|^2 \right) \left\langle f_A \left(|T|^{2\alpha} \right) |T|^{2\alpha} \, x, x \right\rangle \left\langle f_A \left(|T^*|^{2\beta} \right) |T^*|^{2\beta} \, y, y \right\rangle$$

for any $x, y \in H$.

Also if we take $f(z) = 1$ in (3.23), then we get Furuta's inequality (F).

Corollary 3.8 (Dragomir [4]) *With the assumptions of Theorem 3.6 we have the norm inequality*

$$\left\| Tf\left(z \, |T|^{\alpha}\right) g\left(u \, |T|^{\beta}\right) |T|^{\alpha+\beta-1} \right\|^{2} \tag{3.24}$$
$$\leq f_{A}\left(|z|^{2}\right) g_{A}\left(|u|^{2}\right) \left\| f_{A}\left(|T|^{2\alpha}\right) |T|^{2\alpha} \right\| \left\| g_{A}\left(|T^{*}|^{2\beta}\right) |T^{*}|^{2\beta} \right\|$$

and the numerical radius inequality

$$w\left(Tf\left(z \, |T|^{\alpha}\right) g\left(u \, |T|^{\beta}\right) |T|^{\alpha+\beta-1}\right) \tag{3.25}$$
$$\leq \frac{1}{2} \left[f_{A}\left(|z|^{2}\right) g_{A}\left(|u|^{2}\right) \right]^{1/2} \left\| f_{A}\left(|T|^{2\alpha}\right) |T|^{2\alpha} + g_{A}\left(|T^{*}|^{2\beta}\right) |T^{*}|^{2\beta} \right\|.$$

Proof The inequality (3.24) follows from (3.18) by taking the supremum over $x, y \in H$ with $\|x\| = \|y\| = 1$.

Now, from (3.18) we also have

$$\left| \left\langle Tf\left(z \, |T|^{\alpha}\right) g\left(u \, |T|^{\beta}\right) |T|^{\alpha+\beta-1} x, x\right\rangle \right|$$
$$\leq \left[f_{A}\left(|z|^{2}\right) g_{A}\left(|u|^{2}\right) \right]^{1/2} \left\langle f_{A}\left(|T|^{2\alpha}\right) |T|^{2\alpha} x, x\right\rangle^{1/2} \left\langle g_{A}\left(|T^{*}|^{2\beta}\right) |T^{*}|^{2\beta} x, x\right\rangle^{1/2}$$
$$\leq \frac{1}{2} \left[f_{A}\left(|z|^{2}\right) g_{A}\left(|u|^{2}\right) \right]^{1/2} \left\langle \left[f_{A}\left(|T|^{2\alpha}\right) |T|^{2\alpha} + g_{A}\left(|T^{*}|^{2\beta}\right) |T^{*}|^{2\beta} \right] x, x\right\rangle$$

for any $x \in H$.

Taking the supremum over $x \in H$ with $\|x\| = 1$ we get the desired result (3.25). ∎

Remark 3.9 Special cases of (3.18) can be obtained if we take $\beta = \alpha$ or $\beta = 1 - \alpha$ with $\alpha \in [0, 1]$ as in Remarks 3.4 and 3.5. The details are omitted.

3.3 Some Examples

Utilising the inequality (3.8) and the power series representations of some elementary functions as above, we have

$$\left| \left\langle T\left(1_{H} \pm |T|^{\alpha+\beta}\right)^{-1} |T|^{\alpha+\beta-1} x, y\right\rangle \right|^{2} \tag{3.26}$$
$$\leq \left\langle \left(1_{H} - |T|^{2\alpha}\right)^{-1} |T|^{2\alpha} x, x\right\rangle \left\langle \left(1_{H} - |T^{*}|^{2\beta}\right)^{-1} |T^{*}|^{2\beta} y, y\right\rangle$$

and

$$\left| \left\langle T \ln \left[\left(1_H \pm |T|^{\alpha+\beta} \right)^{-1} \right] |T|^{\alpha+\beta-1} x, y \right\rangle \right|^2 \tag{3.27}$$
$$\leq \left\langle \ln \left[\left(1_H - |T|^{2\alpha} \right)^{-1} \right] |T|^{2\alpha} x, x \right\rangle$$
$$\times \left\langle \ln \left[\left(1_H - |T^*|^{2\beta} \right)^{-1} \right] |T^*|^{2\beta} y, y \right\rangle$$

for $T \in \mathcal{B}(H)$, $\alpha, \beta \geq 0$ with $\alpha + \beta \geq 1$ and such that $\|T\| < 1$ and for any $x, y \in H$.

From (3.8) we also have the inequalities

$$\left| \left\langle T \sin \left(|T|^{\alpha+\beta} \right) |T|^{\alpha+\beta-1} x, y \right\rangle \right|^2 \tag{3.28}$$
$$\leq \left\langle \sinh \left(|T|^{2\alpha} \right) |T|^{2\alpha} x, x \right\rangle \left\langle \sinh \left(|T^*|^{2\beta} \right) |T^*|^{2\beta} y, y \right\rangle,$$

$$\left| \left\langle T \cos \left(|T|^{\alpha+\beta} \right) |T|^{\alpha+\beta-1} x, y \right\rangle \right|^2 \tag{3.29}$$
$$\leq \left\langle \cosh \left(|T|^{2\alpha} \right) |T|^{2\alpha} x, x \right\rangle \left\langle \cosh \left(|T^*|^{2\beta} \right) |T^*|^{2\beta} y, y \right\rangle,$$

and

$$\left| \left\langle T \exp \left(|T|^{\alpha+\beta} \right) |T|^{\alpha+\beta-1} x, y \right\rangle \right|^2 \tag{3.30}$$
$$\leq \left\langle \exp \left(|T|^{2\alpha} \right) |T|^{2\alpha} x, x \right\rangle \left\langle \exp \left(|T^*|^{2\beta} \right) |T^*|^{2\beta} y, y \right\rangle$$

for $T \in \mathcal{B}(H)$, $\alpha, \beta \geq 0$ with $\alpha + \beta \geq 1$ and for any $x, y \in H$.

Utilizing the inequality (3.18) we have

$$\left| \left\langle T \exp \left(z |T|^{\alpha} + u |T|^{\beta} \right) |T|^{\alpha+\beta-1} x, y \right\rangle \right|^2 \tag{3.31}$$
$$\leq \exp \left(|z|^2 + |u|^2 \right) \left\langle \exp \left(|T|^{2\alpha} \right) |T|^{2\alpha} x, x \right\rangle \left\langle \exp \left(|T^*|^{2\beta} \right) |T^*|^{2\beta} y, y \right\rangle$$

and

$$\left| \left\langle T \sin \left(z |T|^{\alpha} \right) \cos \left(u |T|^{\beta} \right) |T|^{\alpha+\beta-1} x, y \right\rangle \right|^2 \tag{3.32}$$
$$\leq \sinh \left(|z|^2 \right) \cosh \left(|u|^2 \right) \left\langle \sinh \left(|T|^{2\alpha} \right) |T|^{2\alpha} x, x \right\rangle \left\langle \cosh \left(|T^*|^{2\beta} \right) |T^*|^{2\beta} y, y \right\rangle$$

for $T \in \mathcal{B}(H)$, $\alpha, \beta \geq 0$ with $\alpha + \beta \geq 1$ and for any $z, u \in \mathbb{C}$, $x, y \in H$.

By the same inequality (3.18) we also get

$$\left| \left\langle T \left(1_H \pm z |T|^{\alpha} \right)^{-1} \left(1_H \pm u |T|^{\beta} \right)^{-1} |T|^{\alpha+\beta-1} x, y \right\rangle \right|^2 \tag{3.33}$$
$$\leq \frac{\left\langle \left(1_H - |T|^{2\alpha} \right)^{-1} |T|^{2\alpha} x, x \right\rangle \left\langle \left(1_H - |T^*|^{2\beta} \right)^{-1} |T^*|^{2\beta} y, y \right\rangle}{\left(1 - |z|^2 \right) \left(1 - |u|^2 \right)}$$

for $T \in \mathcal{B}(H)$, $z, u \in \mathbb{C}$ with $\|T\|, |z|, |u| < 1$, $\alpha, \beta \geq 0$ with $\alpha + \beta \geq 1$ and for any $x, y \in H$.

3.4 More Functional Inequalities

We can state the following corollary of Furuta's inequality (F) for the numerical radius w of an operator $V \in B(H)$, namely $w(V) = \sup_{\|x\|=1} |\langle Vx, x \rangle|$, which satisfies the following basic inequalities

$$\frac{1}{2} \|V\| \leq w(V) \leq \|V\|.$$

Corollary 3.10 *Let $T \in B(H)$ and $\alpha, \beta \geq 0$ with $\alpha + \beta \geq 1$. Then we have*

$$w\left(T |T|^{\alpha+\beta-1}\right) \leq \frac{1}{2} \left\| |T|^{2\alpha} + |T^*|^{2\beta} \right\|. \tag{3.34}$$

In particular, we also have

$$w\left(T |T|^{2\alpha-1}\right) \leq \frac{1}{2} \left\| |T|^{2\alpha} + |T^*|^{2\alpha} \right\|, \tag{3.35}$$

for any $\alpha \geq \frac{1}{2}$ and, as a special case,

$$w\left(T |T|\right) \leq \frac{1}{2} \left\| |T|^2 + |T^*|^2 \right\|. \tag{3.36}$$

Proof We have from (F) for any $x \in H$ that

$$\left| \langle T |T|^{\alpha+\beta-1} x, x \rangle \right| \leq \langle |T|^{2\alpha} x, x \rangle^{1/2} \langle |T^*|^{2\beta} x, x \rangle^{1/2} \tag{3.37}$$

$$\leq \frac{1}{2} \left\langle \left[|T|^{2\alpha} + |T^*|^{2\beta} \right] x, x \right\rangle$$

where $\alpha, \beta \geq 0$ with $\alpha + \beta \geq 1$.

Utilising the inequality in (3.37) and taking the supremum over $x \in H$, $\|x\| = 1$ we get

$$w\left(T |T|^{\alpha+\beta-1}\right) = \sup_{\|x\|=1} \left| \langle T |T|^{\alpha+\beta-1} x, x \rangle \right|$$

$$\leq \frac{1}{2} \sup_{\|x\|=1} \left\langle \left[|T|^{2\alpha} + |T^*|^{2\beta} \right] x, x \right\rangle$$

$$= \frac{1}{2} \left\| |T|^{2\alpha} + |T^*|^{2\beta} \right\|. \qquad \blacksquare$$

Now, by the help of power series $f(z) = \sum_{n=0}^{\infty} a_n z^n$ we also have:

Theorem 3.11 *Let $f(z) = \sum_{n=0}^{\infty} a_n z^n$ and be $g(z) = \sum_{n=0}^{\infty} b_n z^n$ be two functions defined by power series with real coefficients and both of them convergent on the open disk $D(0, R) \subset \mathbb{C}$, $R > 0$. If T is a bounded linear operator on the Hilbert space H and $z, u \in \mathbb{C}$ with the property that*

$$|z|^2, |u|^2, \|T\|^2 < R, \tag{3.38}$$

then we have the inequality

$$\begin{aligned}
&|\langle Tf(z|T|) g(u|T|) x, y\rangle|^2 \\
&\leq f_A\left(|z|^2\right) g_A\left(|u|^2\right) \langle f_A\left(|T|^2\right) x, x\rangle \langle |T^*|^2 g_A\left(|T^*|^2\right) y, y\rangle
\end{aligned} \tag{3.39}$$

for any $x, y \in H$.

Proof From Furuta's inequality (F) we have for any natural numbers $n \geq 0$ and $m \geq 1$ the following power inequality

$$|\langle T|T|^{n+m-1} x, y\rangle| \leq \langle |T|^{2n} x, x\rangle^{1/2} \langle |T^*|^{2m} y, y\rangle^{1/2}, \tag{3.40}$$

where $x, y \in H$.

If we multiply this inequality with the positive quantities $|a_n| |z|^n$ and $|b_{m-1}| |u|^{m-1}$, use the triangle inequality and the Cauchy–Bunyakowsky–Schwarz discrete inequality we have successively:

$$\begin{aligned}
&\left| \sum_{n=0}^{k} \sum_{m=1}^{l} a_n z^n b_{m-1} u^{m-1} \langle T|T|^{n+m-1} x, y\rangle \right| \tag{3.41} \\
&\leq \sum_{n=0}^{k} \sum_{m=1}^{l} |a_n| |z|^n |b_{m-1}| |u|^{m-1} |\langle T|T|^{n+m-1} x, y\rangle| \\
&\leq \sum_{n=0}^{k} |a_n| |z|^n \langle |T|^{2n} x, x\rangle^{1/2} \sum_{m=1}^{l} |b_{m-1}| |u|^{m-1} \langle |T^*|^{2m} y, y\rangle^{1/2} \\
&\leq \left(\sum_{n=0}^{k} |a_n| |z|^{2n} \right)^{1/2} \left(\sum_{n=0}^{k} |a_n| \langle |T|^{2n} x, x\rangle \right)^{1/2} \\
&\quad \times \left(\sum_{m=1}^{l} |b_{m-1}| |u|^{2(m-1)} \right)^{1/2} \left(\sum_{m=1}^{l} |b_{m-1}| \langle |T^*|^{2m} y, y\rangle \right)^{1/2}
\end{aligned}$$

for any $x, y \in H$ and $k \geq 0, l \geq 1$.

Observe also that

$$\sum_{n=0}^{k}\sum_{m=1}^{l} a_n z^n b_{m-1} u^{m-1} \left\langle T\,|T|^{n+m-1}\, x, y\right\rangle \tag{3.42}$$

$$= \left\langle T\left(\sum_{n=0}^{k} a_n z^n\,|T|^n\right)\left(\sum_{m=1}^{l} b_{m-1} u^{m-1}\,|T|^{m-1}\right) x, y\right\rangle$$

for any $x, y \in H$ and $k \geq 0, l \geq 1$.

Making use of (3.41) and (3.42) we get

$$\left|\left\langle T\left(\sum_{n=0}^{k} a_n z^n\,|T|^n\right)\left(\sum_{m=1}^{l} b_{m-1} u^{m-1}\,|T|^{m-1}\right) x, y\right\rangle\right| \tag{3.43}$$

$$\leq \left(\sum_{n=0}^{k} |a_n|\,|z|^{2n}\right)^{1/2}\left\langle\sum_{n=0}^{k} |a_n|\,|T|^{2n}\, x, x\right\rangle^{1/2}$$

$$\times \left(\sum_{m=1}^{l} |b_{m-1}|\,|u|^{2(m-1)}\right)^{1/2}\left\langle |T^*|^2 \sum_{m=1}^{l} |b_{m-1}|\,|T^*|^{2(m-1)}\, y, y\right\rangle^{1/2}$$

for any $x, y \in H$ and $k \geq 0, l \geq 1$.

Due to the assumption (3.38) in the theorem, we have that the series $\sum_{n=0}^{\infty} a_n z^n\,|T|^n$, $\sum_{m=0}^{\infty} b_m u^m\,|T|^m$, $\sum_{n=0}^{\infty} |a_n|\,|T|^{2n}$ and $\sum_{m=0}^{\infty} |b_m|\,|T^*|^{2m}$ are convergent in $B\,(H)$ and the series $\sum_{n=0}^{\infty} |a_n|\,|z|^{2n}$ and $\sum_{m=0}^{\infty} |b_m|\,|u|^{2m}$ are convergent in $\mathbb{R}$ and then, by taking the limit over $k \to \infty$ and $l \to \infty$ in (3.43), we deduce the desired result (3.39).   ∎

Remark 3.12 The above inequality (3.39) can provide various particular instances of interest.

For instance, if we take $g = f$ in Theorem 3.11 then we get

$$\left|\left\langle Tf^2\,(z\,|T|)\, x, y\right\rangle\right| \tag{3.44}$$

$$\leq f_A\left(|z|^2\right)\left\langle f_A\left(|T|^2\right) x, x\right\rangle^{1/2}\left\langle |T^*|^2\, f_A\left(|T^*|^2\right) y, y\right\rangle^{1/2}$$

for any $x, y \in H$.

Also if we take $g\,(z) = 1$ in (3.39), then we get

$$\left|\left\langle Tf\,(z\,|T|)\, x, y\right\rangle\right|^2 \leq f_A\left(|z|^2\right)\left\langle f_A\left(|T|^2\right) x, x\right\rangle\left\langle |T^*|^2\, y, y\right\rangle \tag{3.45}$$

for any $x, y \in H$.

Corollary 3.13 (Dragomir [6]) *With the assumptions of Theorem 3.11 we have the norm inequality*

$$\|Tf(z\,|T|)\,g(u\,|T|)\|^2 \tag{3.46}$$

$$\leq f_A\left(|z|^2\right)g_A\left(|u|^2\right)\left\|f_A\left(|T|^2\right)\right\|\left\||T^*|^2\,g_A\left(|T^*|^2\right)\right\|$$

and the numerical radius inequality

$$w\left(Tf(z\,|T|)\,g(u\,|T|)\right) \tag{3.47}$$

$$\leq \frac{1}{2}\left[f_A\left(|z|^2\right)g_A\left(|u|^2\right)\right]^{1/2}\left\|f_A\left(|T|^2\right)+|T^*|^2\,g_A\left(|T^*|^2\right)\right\|.$$

Proof The inequality (3.46) follows from (3.39) by taking the supremum over $x, y \in H$ with $\|x\| = \|y\| = 1$.

From (3.39) we also have the inequality

$$|\langle Tf(z\,|T|)\,g(u\,|T|)\,x, x\rangle|$$

$$\leq \left[f_A\left(|z|^2\right)g_A\left(|u|^2\right)\right]^{1/2}\left\langle f_A\left(|T|^2\right)x, x\right\rangle^{1/2}\left\langle |T^*|^2\,g_A\left(|T^*|^2\right)x, x\right\rangle^{1/2}$$

$$\leq \frac{1}{2}\left[f_A\left(|z|^2\right)g_A\left(|u|^2\right)\right]^{1/2}\left\langle\left[f_A\left(|T|^2\right)+|T^*|^2\,g_A\left(|T^*|^2\right)\right]x, x\right\rangle^{1/2}$$

for any $x \in H$, which, by taking the supremum over $\|x\| = 1$ produces the desired result (3.47). ∎

The following result also holds:

Theorem 3.14 (Dragomir [6]) *Let $f(z) = \sum_{n=0}^{\infty} a_n z^n$ be a function defined by power series with real coefficients and convergent on the open disk $D(0, R) \subset \mathbb{C}$, $R > 0$. If T is a bounded linear operator on the Hilbert space H with the property that $\|T\|^2 < R$, then we have the inequality*

$$\left|\langle T\,|T|\,f\left(|T|^2\right)x, y\rangle\right|^2 \leq \langle |T|^2\,f_A\left(|T|^2\right)x, x\rangle\left\langle |T^*|^2\,f_A\left(|T^*|^2\right)y, y\right\rangle \tag{3.48}$$

for any $x, y \in H$.

Proof From Furuta's inequality (F) we have for any natural numbers $n \geq 1$ the power inequality

$$\left|\langle T\,|T|^{2n-1}\,x, y\rangle\right| \leq \langle |T|^{2n}\,x, x\rangle^{1/2}\left\langle |T^*|^{2n}\,y, y\right\rangle^{1/2} \tag{3.49}$$

where $x, y \in H$.

If we multiply this inequality with the positive quantities $|a_{n-1}|$, use the triangle inequality and the Cauchy–Bunyakowsky–Schwarz discrete inequality we have successively

$$\left| \left\langle \sum_{n=1}^{k} a_{n-1} T \, |T|^{2n-1} x, y \right\rangle \right| \tag{3.50}$$

$$\leq \sum_{n=1}^{k} |a_{n-1}| \left| \langle T \, |T|^{2n-1} x, y \rangle \right|$$

$$\leq \sum_{n=1}^{k} |a_{n-1}| \left\langle |T|^{2n} x, x \right\rangle^{1/2} \left\langle |T^*|^{2n} y, y \right\rangle^{1/2}$$

$$\leq \left\langle \sum_{n=1}^{k} |a_{n-1}| \, |T|^{2n} x, x \right\rangle^{1/2} \left\langle \sum_{n=1}^{k} |a_{n-1}| \, |T^*|^{2n} y, y \right\rangle^{1/2}$$

for any $x, y \in H$ and $k \geq 1$.

Observe also that

$$\sum_{n=1}^{k} a_{n-1} T \, |T|^{2n-1} = T \, |T| \sum_{n=1}^{k} a_{n-1} |T|^{2(n-1)},$$

$$\sum_{n=1}^{k} |a_{n-1}| \, |T|^{2n} = |T|^2 \sum_{n=1}^{k} |a_{n-1}| \, |T|^{2(n-1)}$$

and

$$\sum_{n=1}^{k} |a_{n-1}| \, |T^*|^{2n} = |T^*|^2 \sum_{n=1}^{k} |a_{n-1}| \, |T^*|^{2(n-1)}$$

for any $k \geq 1$.

Therefore, by (3.50) we have the inequality

$$\left\langle T \, |T| \sum_{n=1}^{k} a_{n-1} |T|^{2(n-1)} x, y \right\rangle^2 \tag{3.51}$$

$$\leq \left\langle |T|^2 \sum_{n=1}^{k} |a_{n-1}| \, |T|^{2(n-1)} x, x \right\rangle \left\langle |T^*|^2 \sum_{n=1}^{k} |a_{n-1}| \, |T^*|^{2(n-1)} y, y \right\rangle$$

for any $x, y \in H$ and $k \geq 1$.

Due to the assumption $\|T\|^2 < R$, we have that the series $\sum_{n=0}^{\infty} a_n |T|^{2n}$, $\sum_{n=0}^{\infty} |a_n| \, |T|^{2n}$ and $\sum_{n=0}^{\infty} |a_n| \, |T^*|^{2n}$ are convergent in $B(H)$ and taking the limit over $k \to \infty$ in (3.51) we deduce the desired result from (3.48). ∎

Corollary 3.15 *With the assumptions of Theorem 3.14 we have the norm inequality*

$$\left\| T \, |T| \, f\left(|T|^2 \right) \right\|^2 \leq \left\| |T|^2 \, f_A\left(|T|^2 \right) \right\| \left\| |T^*|^2 \, f_A\left(|T^*|^2 \right) \right\|$$

and the numerical radius inequality

$$w\left(T\,|T|\,f\left(|T|^2\right)\right) \le \frac{1}{2}\left\| |T|^2\,f_A\left(|T|^2\right) + |T^*|^2\,f_A\left(|T^*|^2\right)\right\|.$$

The following result for functions of normal operators holds.

Theorem 3.16 (Dragomir [6]) *Let* $f(z) = \sum_{n=0}^{\infty} a_n z^n$ *be a function defined by power series with real coefficients and convergent on the open disk* $D(0, R) \subset \mathbb{C}$, $R > 0$. *If* N *is a normal operator on the Hilbert space* H *and* $\alpha, \beta \ge 0$ *with* $\alpha + \beta \ge 1$ *with the property that* $\|N\|^{2\alpha}$, $\|N\|^{2\beta} < R$, *then we have the inequality*

$$\left|\left\langle f\left(N\,|N|^{(\alpha+\beta-1)}\right)x, y\right\rangle\right|^2 \le \left\langle f_A\left(|N|^{2\alpha}\right)x, x\right\rangle\left\langle f_A\left(|N|^{2\beta}\right)y, y\right\rangle \quad (3.52)$$

for any $x, y \in H$.

Proof Utilising Furuta's inequality written for N^n we have

$$\left|\left\langle N^n\,|N^n|^{\alpha+\beta-1}x, y\right\rangle\right|^2 \le \left\langle |N^n|^{2\alpha}x, x\right\rangle\left\langle |(N^n)^*|^{2\beta}y, y\right\rangle \quad (3.53)$$

for any $x, y \in H$.

Since N is normal, then

$$\begin{aligned}
|N^n|^2 &= (N^n)^* N^n = N^* \cdots N^* N \cdots N \\
&= N^* \cdots N N^* \cdots N = \cdots \\
&= (N^*N) \cdots (N^*N) = |N|^{2n}
\end{aligned}$$

for any natural number n, and, similarly,

$$\left|(N^n)^*\right|^2 = \left|(N^*)^n\right|^2 = \left|N^*\right|^{2n} = |N|^{2n}$$

for any $n \in \mathbb{N}$.

These imply that $|N^n|^{2\alpha} = |N|^{2\alpha n}$, $|(N^n)^*|^{2\beta} = |N|^{2\beta n}$ and $|N^n|^{\alpha+\beta-1} = |N|^{(\alpha+\beta-1)n}$ for any $\alpha, \beta \ge 0$ and for any $n \in \mathbb{N}$.

Utilising the spectral representation for Borel functions of normal operators on Hilbert spaces, see for instance [1, p. 67], we have for any $\alpha, \beta \ge 0$ and for any $n \in \mathbb{N}$ that

$$\begin{aligned}
N^n\,|N|^{(\alpha+\beta-1)n} &= \int_{\sigma(N)} z^n\,|z|^{(\alpha+\beta-1)n}\,dP(z) \\
&= \int_{\sigma(N)} \left[z\,|z|^{(\alpha+\beta-1)}\right]^n\,dP(z) \\
&= \left[N\,|N|^{(\alpha+\beta-1)}\right]^n,
\end{aligned}$$

where P is the spectral measure associated to the operator N and $\sigma(N)$ is its spectrum.

Therefore, the inequality (3.53) can be written as

$$\left|\left\langle\left[N\,|N|^{(\alpha+\beta-1)}\right]^n x, y\right\rangle\right| \le \left\langle\left[|N|^{2\alpha}\right]^n x, x\right\rangle^{1/2}\left\langle\left[|N|^{2\beta}\right]^n y, y\right\rangle^{1/2} \qquad (3.54)$$

for any $x, y \in H$ and for any $n \in \mathbb{N}$.

If we multiply the inequality (3.54) by $|a_n| \ge 0$, sum over n from 0 to $k \ge 1$ and utilize the Cauchy–Bunyakowsky–Schwarz discrete inequality, we have successively

$$\left|\left\langle \sum_{n=0}^{k} a_n\left[N\,|N|^{(\alpha+\beta-1)}\right]^n x, y\right\rangle\right| \qquad (3.55)$$

$$\le \sum_{n=0}^{k} |a_n|\left|\left\langle\left[N\,|N|^{(\alpha+\beta-1)}\right]^n x, y\right\rangle\right|$$

$$\le \sum_{n=0}^{k} |a_n|\left\langle\left[|N|^{2\alpha}\right]^n x, x\right\rangle^{1/2}\left\langle\left[|N|^{2\beta}\right]^n y, y\right\rangle^{1/2}$$

$$\le \left(\sum_{n=0}^{k} |a_n|\left[|N|^{2\alpha}\right]^n x, x\right)^{1/2}\left(\sum_{n=0}^{k} |a_n|\left[|N|^{2\beta}\right]^n y, y\right)^{1/2}$$

for any $x, y \in H$ and for any $k \ge 1$.

Since $\|N\|^{2\alpha}, \|N\|^{2\beta} < R$ then $\left\|N\,|N|^{(\alpha+\beta-1)}\right\| < R$ and the series

$$\sum_{n=0}^{\infty} |a_n|\left[|N|^{2\alpha}\right]^n, \sum_{n=0}^{\infty} |a_n|\left[|N|^{2\beta}\right]^n$$

and

$$\sum_{n=0}^{\infty} a_n\left[N\,|N|^{(\alpha+\beta-1)}\right]^n$$

are convergent in the Banach algebra $B(H)$.

Taking the limit over $k \to \infty$ in the inequality (3.55) we deduce the desired result from (3.52).   ■

Corollary 3.17 (Dragomir [6]) *With the assumptions of Theorem 3.16, we have the inequality*

$$\left\|f\left(N\,|N|^{(\alpha+\beta-1)}\right)\right\|^2 \le \left\|f_A\left(|N|^{2\alpha}\right)\right\|\left\|f_A\left(|N|^{2\beta}\right)\right\|. \qquad (3.56)$$

Remark 3.18 If we take $\beta = 1 - \alpha$ with $\alpha \in [0, 1]$ in (3.52), then we get the following generalization of Kato's inequality for normal operators

$$|\langle f(N) x, y\rangle|^2 \le \langle f_A (|N|^{2\alpha}) x, x\rangle \langle f_A (|N|^{2(1-\alpha)}) y, y\rangle \qquad (3.57)$$

where $x, y \in H$ and $\|N\|^{2\alpha}, \|N\|^{2(1-\alpha)} < R$.

3.5 Applications for Some Elementary Functions

We can give now some examples:

Example 3.19 Let $x, y \in H$.
   (a) If we take $f(z) = \sin z$ and $g(z) = \cos z$ in (3.39), then we get

$$|\langle T \sin(z\,|T|) \cos(u\,|T|) x, y\rangle|^2 \qquad (3.58)$$
$$\le \sinh(|z|^2) \cosh(|u|^2)$$
$$\times \langle \sinh(|T|^2) x, x\rangle \langle |T^*|^2 \cosh(|T^*|^2) y, y\rangle$$

for any $z \in \mathbb{C}$ and $T \in B(H)$.
   (b) If we take $f(z) = \ln \frac{1}{1+z}$ and $g(z) = \ln \frac{1}{1-z}$ in (3.39), then we get

$$\left|\langle T \ln(1_H + z\,|T|)^{-1} \ln(1_H - z\,|T|)^{-1} x, y\rangle\right|^2 \qquad (3.59)$$
$$\le \left(\ln \frac{1}{1-|z|^2}\right)^2$$
$$\times \left\langle \ln\left(1_H - |T|^2\right)^{-1} x, x\right\rangle \left\langle |T^*|^2 \ln\left(1_H - |T^*|^2\right)^{-1} y, y\right\rangle$$

for any $z \in \mathbb{C}$ and $T \in B(H)$ with $|z| < 1$ and $\|T\| < 1$.
   (c) If we take $f(z) = \exp(z)$ and $g(z) = \exp(z)$ in (3.39), then we get

$$|\langle T \exp[(z+u)\,|T|] x, y\rangle|^2 \qquad (3.60)$$
$$\le \exp(|z|^2) \exp(|u|^2)$$
$$\times \langle \exp(|T|^2) x, x\rangle \langle |T^*|^2 \exp(|T^*|^2) y, y\rangle$$

for any $z, u \in \mathbb{C}$ and $T \in B(H)$.
   (d) By the inequality (3.45) we have

$$\left|\langle T \sin^{-1}(z\,|T|) x, y\rangle\right|^2 \le \sin^{-1}(|z|^2) \langle \sin^{-1}(|T|^2) x, x\rangle \langle |T^*|^2 y, y\rangle \qquad (3.61)$$

and

$$\left|\left\langle T\tanh^{-1}\left(z\left|T\right|\right)x,y\right\rangle\right|^2 \tag{3.62}$$
$$\leq \tanh^{-1}\left(|z|^2\right)\left\langle\tanh^{-1}\left(|T|^2\right)x,x\right\rangle\left\langle\left|T^*\right|^2 y,y\right\rangle$$

for any $z\in\mathbb{C}$ and $T\in B(H)$ with $|z|<1$ and $\|T\|<1$.

Example 3.20 Let $x,y\in H$.
(a) If we take $f(z)=\frac{1}{1\pm z}$ in (3.48), then we get

$$\left|\left\langle T\left|T\right|\left(1_H\pm|T|^2\right)^{-1}x,y\right\rangle\right|^2 \tag{3.63}$$
$$\leq\left\langle|T|^2\left(1_H-|T|^2\right)^{-1}x,x\right\rangle\left\langle\left|T^*\right|^2\left(1_H-\left|T^*\right|^2\right)^{-1}y,y\right\rangle$$

for any $T\in B(H)$ with $\|T\|<1$.
(b) If we take $f(z)=\ln\frac{1}{1\pm z}$ in (3.48), then we get

$$\left|\left\langle T\left|T\right|\ln\left(1_H\pm|T|^2\right)^{-1}x,y\right\rangle\right|^2 \tag{3.64}$$
$$\leq\left\langle|T|^2\ln\left(1_H-|T|^2\right)^{-1}x,x\right\rangle\left\langle\left|T^*\right|^2\ln\left(1_H-\left|T^*\right|^2\right)^{-1}y,y\right\rangle$$

for any $T\in B(H)$ with $\|T\|<1$.
(c) If we take $f(z)=\exp(z)$ in (3.48), then we get

$$\left|\left\langle T\left|T\right|\exp\left(|T|^2\right)x,y\right\rangle\right|^2 \tag{3.65}$$
$$\leq\left\langle|T|^2\exp\left(|T|^2\right)x,x\right\rangle\left\langle\left|T^*\right|^2\exp\left(\left|T^*\right|^2\right)y,y\right\rangle$$

for any $T\in B(H)$.

Example 3.21 Let N be a normal operator on the Hilbert space H, $\alpha,\beta\geq 0$ with $\alpha+\beta\geq 1$ and $x,y\in H$.
(a) If we take $f(z)=\frac{1}{1\pm z}$ in (3.52), then we get

$$\left|\left\langle\left(1_H\pm N\left|N\right|^{(\alpha+\beta-1)}\right)^{-1}x,y\right\rangle\right|^2 \tag{3.66}$$
$$\leq\left\langle\left(1_H-|N|^{2\alpha}\right)^{-1}x,x\right\rangle\left\langle\left(1_H-|N|^{2\beta}\right)^{-1}y,y\right\rangle$$

provided $\|N\|<1$.
In particular, we have

$$\left|\left\langle(1_H\pm N)^{-1}x,y\right\rangle\right|^2 \tag{3.67}$$
$$\leq\left\langle\left(1_H-|N|^{2\alpha}\right)^{-1}x,x\right\rangle\left\langle\left(1_H-|N|^{2(1-\alpha)}\right)^{-1}y,y\right\rangle,$$

for $\alpha \in [0, 1]$.

(b) If we take $f(z) = \exp(z)$ in (3.52), then we get

$$\left|\left\langle \exp\left(N \,|N|^{(\alpha+\beta-1)}\right) x, y\right\rangle\right|^2 \leq \left\langle \exp\left(|N|^{2\alpha}\right) x, x\right\rangle \left\langle \exp\left(|N|^{2\beta}\right) y, y\right\rangle. \tag{3.68}$$

As a special case, we have

$$\left|\left\langle \exp(N) x, y\right\rangle\right|^2 \leq \left\langle \exp\left(|N|^{2\alpha}\right) x, x\right\rangle \left\langle \exp\left(|N|^{2(1-\alpha)}\right) y, y\right\rangle, \tag{3.69}$$

for $\alpha \in [0, 1]$.

3.6 General Vector Inequalities

The following result provides a general extension for four operators of the Schwarz inequality:

Theorem 3.22 (Dragomir [7]) *Let $A, B, C, D \in \mathcal{B}(H)$. Then for $x, y \in H$ we have the inequality*

$$\left|\langle DCBAx, y\rangle\right|^2 \leq \left\langle A^* \,|B|^2 \,Ax, x\right\rangle \left\langle D \,|C^*|^2 \,D^* y, y\right\rangle. \tag{3.70}$$

The equality case holds in (3.70) if the vectors BAx and C^*D^*y are linearly dependent in H.

Proof The Schwarz inequality in the Hilbert space H states that for any $u, v \in H$ we have the inequality

$$|\langle u, v\rangle|^2 \leq \|u\|^2 \|v\|^2 \tag{3.71}$$

with equality if and only if the vectors u and v are linearly dependent in H.

Now, if we take $u = BAx$ and $v = C^*D^*y$ then we have

$$\|u\|^2 = \langle BAx, BAx\rangle = \left\langle B^*BAx, Ax\right\rangle$$
$$= \left\langle A^*B^*BAx, x\right\rangle = \left\langle A^* \,|B|^2 \,Ax, x\right\rangle,$$

$$\|v\|^2 = \left\langle C^*D^*y, C^*D^*y\right\rangle = \left\langle CC^*D^*y, D^*y\right\rangle$$
$$= \left\langle DCC^*D^*y, y\right\rangle = \left\langle D \,|C^*|^2 \,D^*y, y\right\rangle$$

and

$$\langle u, v\rangle = \left\langle BAx, C^*D^*y\right\rangle = \left\langle CBAx, D^*y\right\rangle = \langle DCBAx, y\rangle.$$

Utilising (3.71) we deduce the desired result (3.70).   ∎

Corollary 3.23 (Dragomir [7]) *The Furuta inequality (F) for α, $\beta \geq 0$ with $\alpha + \beta \geq$ 1 is a particular case of (3.70).*

Proof Let $T = U\,|T|$ be the polar decomposition of an operator T, where U is partial isometry and the kernel $N\,(U) = N\,(|T|)$.

If we take $D = U, C = |T|^\beta$, $B = 1_H$ and $A = |T|^\alpha$ then we have

$$DCBA = U\,|T|^\beta\,|T|^\alpha = U\,|T|\,|T|^{\alpha+\beta-1} = T\,|T|^{\alpha+\beta-1},$$

$$A^*\,|B|^2\,A = |T|^\alpha\,|T|^\alpha = |T|^{2\alpha}$$

and

$$D\,\left|C^*\right|^2 D^* = U\,|T|^{2\beta}\,U^* = \left|T^*\right|^{2\beta},$$

which by (3.70) implies the desired inequality (F).   ∎

The following similar result also holds

Corollary 3.24 (Dragomir [7]) *For any operator $T \in \mathcal{B}\,(H)$ and any α, $\beta \geq 1$ we have the inequality*

$$\left|\langle T\,|T|^{\beta-1}\,T\,|T|^{\alpha-1}\,x, y\rangle\right|^2 \leq \langle |T|^{2\alpha}\,x, x\rangle\left\langle \left|T^*\right|^{2\beta}\,y, y\right\rangle, \qquad (3.72)$$

where $x, y \in H$.

Proof Let $T = U\,|T|$ be the polar decomposition of an operator T, where U is partial isometry and the kernel $N\,(U) = N\,(|T|)$.

If we take $D = U, C = |T|^\beta$, $B = U$ and $A = |T|^\alpha$ then we have

$$DCBA = U\,|T|^\beta\,U\,|T|^\alpha = U\,|T|\,|T|^{\beta-1}\,U\,|T|\,|T|^{\alpha-1} = T\,|T|^{\beta-1}\,T\,|T|^{\alpha-1},$$

$$A^*\,|B|^2\,A = |T|^\alpha\,U^*U\,|T|^\alpha = |T|^{\alpha-1}\,|T|\,U^*U\,|T|\,|T|^{\alpha-1}$$
$$= |T|^{\alpha-1}\,T^*T\,|T|^{\alpha-1} = |T|^{\alpha-1}\,|T|^2\,|T|^{\alpha-1} = |T|^{2\alpha}$$

and

$$D\,\left|C^*\right|^2 D^* = U\,|T|^{2\beta}\,U^* = \left|T^*\right|^{2\beta},$$

which by (3.70) implies the desired inequality (3.72).   ∎

Remark 3.25 The above inequality (3.72) contains some nice particular inequalities as follows:

$$\left|\left\langle \left(T \, |T|^{\alpha-1}\right)^2 x, y \right\rangle\right|^2 \le \left\langle |T|^{2\alpha} x, x \right\rangle \left\langle \left|T^*\right|^{2\alpha} y, y \right\rangle, \tag{3.73}$$

for $\alpha \ge 1$ producing for $\alpha = 1$ the result

$$\left|\left\langle T^2 x, y \right\rangle\right|^2 \le \left\langle |T|^2 x, x \right\rangle \left\langle \left|T^*\right|^2 y, y \right\rangle, \tag{3.74}$$

and for $\alpha = 2$ the result

$$\left|\left\langle (T \, |T|)^2 x, y \right\rangle\right|^2 \le \left\langle |T|^4 x, x \right\rangle \left\langle \left|T^*\right|^4 y, y \right\rangle, \tag{3.75}$$

for any $x, y \in H$.

If we take $\alpha = 1$ in (3.72), then we get

$$\left|\left\langle T \, |T|^{\beta-1} \, T x, y \right\rangle\right|^2 \le \left\langle |T|^2 x, x \right\rangle \left\langle \left|T^*\right|^{2\beta} y, y \right\rangle, \tag{3.76}$$

for any $\beta \ge 1$ and if we take $\beta = 1$ then we also get

$$\left|\left\langle T^2 \, |T|^{\alpha-1} x, y \right\rangle\right|^2 \le \left\langle |T|^{2\alpha} x, x \right\rangle \left\langle \left|T^*\right|^2 y, y \right\rangle, \tag{3.77}$$

for any $x, y \in H$.

Corollary 3.26 (Dragomir [7]) *For any operator $T \in \mathcal{B}(H)$ and any $\gamma, \delta \ge 0$ we have the inequality*

$$\left|\left\langle |T|^\gamma \, T^2 \, |T|^\delta x, y \right\rangle\right|^2 \le \left\langle |T|^{2\delta+2} x, x \right\rangle \left\langle \left|T^* \, |T|^\gamma\right|^2 y, y \right\rangle, \tag{3.78}$$

where $x, y \in H$.

Proof If we take $D = |T|^\gamma$, $C = T$, $B = T$ and $A = |T|^\delta$ then we have

$$DCBA = |T|^\gamma \, T^2 \, |T|^\delta,$$

$$A^* |B|^2 A = |T|^\delta \, |T|^2 \, |T|^\delta = |T|^{2\delta+2}$$

and

$$D \left|C^*\right|^2 D^* = |T|^\gamma \left|T^*\right|^2 |T|^\gamma = |T|^\gamma \, T T^* \, |T|^\gamma$$
$$= |T|^\gamma \, T \left(|T|^\gamma \, T\right)^* = \left|\left(|T|^\gamma \, T\right)^*\right|^2 = \left|T^* \, |T|^\gamma\right|^2,$$

which by (3.70) implies the desired inequality (3.78). ∎

Remark 3.27 The particular case $\gamma = \delta = 1$ provides the inequality

$$\left|\langle |T| \, T^2 \, |T| \, x, y \rangle\right|^2 \leq \langle |T|^4 x, x \rangle \left\langle \left|T^* \, |T|\right|^2 y, y \right\rangle, \qquad (3.79)$$

for any $x, y \in H$.

We also have

Corollary 3.28 (Dragomir [7]) *For any operator $T \in \mathcal{B}(H)$ and any $\gamma, \delta \geq 0$ we have the inequalities*

$$\left|\langle |T|^{\gamma+\delta+2} x, y \rangle\right|^2 \leq \langle |T|^{2\delta+2} x, x \rangle \langle |T|^{2\gamma+2} y, y \rangle \qquad (3.80)$$

and

$$\left|\left\langle |T|^{\gamma} \, |T^*|^2 \, |T|^{\delta} x, y \right\rangle\right|^2 \leq \left\langle \left|T^* \, |T|^{\delta}\right|^2 x, x \right\rangle \left\langle \left|T^* \, |T|^{\gamma}\right|^2 y, y \right\rangle \qquad (3.81)$$

where $x, y \in H$.

Proof If we take $D = |T|^{\gamma}$, $C = T^*$, $B = T$ and $A = |T|^{\delta}$ then we have

$$DCBA = |T|^{\gamma} \, T^* T \, |T|^{\delta} = |T|^{\gamma} \, |T|^2 \, |T|^{\delta} = |T|^{\gamma+\delta+2},$$

$$A^* \, |B|^2 \, A = |T|^{\delta} \, |T|^2 \, |T|^{\delta} = |T|^{2\delta+2}$$

and

$$D \, |C^*|^2 \, D^* = |T|^{\gamma} \, |T|^2 \, |T|^{\gamma} = |T|^{2\gamma+2}$$

which by (3.70) implies the desired inequality (3.80).

The dual choice $D = |T|^{\gamma}$, $C = T$, $B = T^*$ and $A = |T|^{\delta}$ gives

$$DCBA = |T|^{\gamma} \, |T^*|^2 \, |T|^{\delta},$$

$$A^* \, |B|^2 \, A = |T|^{\delta} \, |T^*|^2 \, |T|^{\delta} = \left|T^* \, |T|^{\delta}\right|^2$$

and

$$D \, |C^*|^2 \, D^* = |T|^{\gamma} \, |T^*|^2 \, |T|^{\gamma} = \left|T^* \, |T|^{\gamma}\right|^2,$$

which by (3.70) produces (3.81). ∎

Remark 3.29 If we take $\delta = \gamma$ in (3.81), then we get

$$\left|\left\langle |T|^{\gamma} \, |T^*|^2 \, |T|^{\gamma} x, y \right\rangle\right|^2 \leq \left\langle \left|T^* \, |T|^{\gamma}\right|^2 x, x \right\rangle \left\langle \left|T^* \, |T|^{\gamma}\right|^2 y, y \right\rangle \qquad (3.82)$$

where $x, y \in H$.

The following corollary also holds

Corollary 3.30 (Dragomir [7]) *For any operator* $T \in \mathcal{B}(H)$ *and any* $\beta \geq 0$ *we have the inequalities*

$$\left| \left\langle T \left| T^* \right|^\beta T x, y \right\rangle \right|^2 \leq \left\langle |T|^2 x, x \right\rangle \left\langle T \left| T^* \right|^{2\beta} T^* y, y \right\rangle \qquad (3.83)$$

and

$$\left| \left\langle T |T|^\beta T x, y \right\rangle \right|^2 \leq \left\langle |T|^2 x, x \right\rangle \left\langle T |T|^{2\beta} T^* y, y \right\rangle \qquad (3.84)$$

where $x, y \in H$.

Proof Let $T = U |T|$ be the polar decomposition of an operator T, where U is partial isometry and the kernel $N(U) = N(|T|)$.

If we take $D = U, C = |T| \, |T^*|^\beta$, $B = U$ and $A = |T|$ then we have

$$DCBA = U |T| \, \big| T^* \big|^\beta U |T| = T \, \big| T^* \big|^\beta T,$$

$$A^* |B|^2 A = |T| \, U^* U |T| = T^* T = |T|^2$$

and

$$D \left| C^* \right|^2 D^* = UCC^* U^* = U |T| \, \big| T^* \big|^\beta \, \big| T^* \big|^\beta |T| \, U^*$$
$$= T \, \big| T^* \big|^{2\beta} T^*,$$

which by (3.70) produces (3.83).

Now, if we take $D = U, C = |T|^{\beta+1}$, $B = U$ and $A = |T|$ then we have

$$DCBA = U |T|^{\beta+1} U |T| = T |T|^\beta T,$$

$$A^* |B|^2 A = |T|^2$$

and

$$D \left| C^* \right|^2 D^* = UCC^* U^* = U |T|^{\beta+1} |T|^{\beta+1} U^* = T |T|^{2\beta} T^*.$$

∎

Remark 3.31 The case $\beta = 1$ produces from the inequalities (3.83) and (3.84) the simple results

$$\left| \left\langle T \left| T^* \right| T x, y \right\rangle \right|^2 \leq \left\langle |T|^2 x, x \right\rangle \left\langle T^2 \left(T^* \right)^2 y, y \right\rangle \qquad (3.85)$$

and

$$|\langle T\,|T|\,Tx, y\rangle|^2 \le \langle |T|^2 x, x\rangle \langle |T^*|^4 y, y\rangle \tag{3.86}$$

for any $x, y \in H$.

3.7 Norm and Numerical Radius Inequalities

We can state the following corollary of Furuta's inequality for the numerical radius w of an operator $V \in \mathcal{B}(H)$, namely $w(V) = \sup_{\|x\|=1} |\langle Vx, x\rangle|$, which satisfies the following basic inequalities

$$\frac{1}{2}\|V\| \le w(V) \le \|V\|. \tag{3.87}$$

Theorem 3.32 (Dragomir [7]) *Let* $A, B, C, D \in \mathcal{B}(H)$. *Then we have*

$$\|DCBA\|^2 \le \left\| A^* |B|^2 A \right\| \left\| D\,|C^*|^2\,D^* \right\| \tag{3.88}$$

and for any $r \ge 1$

$$w^r(DCBA) \le \frac{1}{2} \left\| \left(A^* |B|^2 A \right)^r + \left(D\,|C^*|^2\,D^* \right)^r \right\|. \tag{3.89}$$

Proof Taking the supremum over $x, y \in H$ with $\|x\| = \|y\| = 1$ in (3.70) we have

$$\begin{aligned}
\|DCBA\|^2 &= \sup_{\|x\|=\|y\|=1} |\langle DCBAx, y\rangle|^2 \\
&\le \sup_{\|x\|=\|y\|=1} \left[\langle A^* |B|^2 Ax, x\rangle \langle D\,|C^*|^2\,D^* y, y\rangle \right] \\
&= \sup_{\|x\|=1} \langle A^* |B|^2 Ax, x\rangle \sup_{\|y\|=1} \langle D\,|C^*|^2\,D^* y, y\rangle \\
&= \left\| A^* |B|^2 A \right\| \left\| D\,|C^*|^2\,D^* \right\|
\end{aligned}$$

and the inequality (3.88) is proved.

By taking $x = y$ in (3.70) and utilising the increasing monotonicity of the power means for two positive numbers, we have for any $r \ge 1$ that

$$|\langle DCBAx, x\rangle| \leq \left[\langle A^* |B|^2 Ax, x\rangle \langle D |C^*|^2 D^*x, x\rangle\right]^{1/2} \qquad (3.90)$$

$$\leq \frac{\langle A^* |B|^2 Ax, x\rangle + \langle D |C^*|^2 D^*x, x\rangle}{2}$$

$$\leq \left[\frac{\langle A^* |B|^2 Ax, x\rangle^r + \langle D |C^*|^2 D^*x, x\rangle^r}{2}\right]^{1/r}$$

for any $x \in H$.

Now, utilising Hölder-McCarthy inequality $\langle Px, x\rangle^r \leq \langle P^r x, x\rangle$, $x \in H$, $\|x\| = 1$ that holds for any positive operator P and any power $r \geq 1$ we have

$$\frac{\langle A^* |B|^2 Ax, x\rangle^r + \langle D |C^*|^2 D^*x, x\rangle^r}{2} \qquad (3.91)$$

$$\leq \frac{\langle (A^* |B|^2 A)^r x, x\rangle + \langle (D |C^*|^2 D^*)^r x, x\rangle}{2}$$

$$= \left\langle \frac{(A^* |B|^2 A)^r + (D |C^*|^2 D^*)^r}{2} x, x\right\rangle$$

for any $x \in H$, $\|x\| = 1$.

By making use of (3.90) and (3.91) we get the inequality of interest

$$|\langle DCBAx, x\rangle|^r \leq \left\langle \frac{(A^* |B|^2 A)^r + (D |C^*|^2 D^*)^r}{2} x, x\right\rangle \qquad (3.92)$$

for any $x \in H$, $\|x\| = 1$.

Finally, by taking the supremum over $x \in H$, $\|x\| = 1$ in (3.92) we deduce the desired result (3.89). ∎

The above theorem has a number of particular cases for one operator that are of interest:

Corollary 3.33 (Dragomir [7]) *1. Let $T \in B(H)$, $r \geq 1$ and $\alpha, \beta \geq 0$ with $\alpha + \beta \geq 1$. Then we have*

$$w^r \left(T |T|^{\alpha+\beta-1}\right) \leq \frac{1}{2} \left\| |T|^{2\alpha r} + |T^*|^{2\beta r} \right\|. \qquad (3.93)$$

In particular, we also have

$$w^r \left(T |T|^{2\alpha-1}\right) \leq \frac{1}{2} \left\| |T|^{2\alpha r} + |T^*|^{2\alpha r} \right\|, \qquad (3.94)$$

for any $\alpha \geq \frac{1}{2}$ and

$$w^r \left(T |T|\right) \leq \frac{1}{2} \left\| |T|^{2r} + |T^*|^{2r} \right\|. \qquad (3.95)$$

2. *For any operator* $T \in \mathcal{B}(H)$, $r \geq 1$ *and any* α, $\beta \geq 1$ *we have the inequality*

$$w^r \left(T |T|^{\beta-1} T |T|^{\alpha-1} \right) \leq \frac{1}{2} \left\| |T|^{2\alpha r} + |T^*|^{2\beta r} \right\|. \tag{3.96}$$

In particular, we also have

$$w^r \left(\left(T |T|^{\alpha-1} \right)^2 \right) \leq \frac{1}{2} \left\| |T|^{2\alpha r} + |T^*|^{2\alpha r} \right\|, \tag{3.97}$$

for any $\alpha \geq 1$ *which provides the result*

$$w^r \left(T^2 \right) \leq \frac{1}{2} \left\| |T|^{2r} + |T^*|^{2r} \right\|. \tag{3.98}$$

3. *For any operator* $T \in \mathcal{B}(H)$, $r \geq 1$ *and any* $\beta \geq 0$ *we have the inequalities*

$$w^r \left(T |T^*|^{\beta} T \right) \leq \frac{1}{2} \left\| |T|^{2r} + \left[T |T^*|^{2\beta} T^* \right]^r \right\| \tag{3.99}$$

and

$$w^r \left(T |T|^{\beta} T \right) \leq \frac{1}{2} \left\| |T|^{2r} + \left[T |T^*|^{2\beta} T^* \right]^r \right\|. \tag{3.100}$$

In particular, we have

$$w^r \left(T |T^*| T \right) \leq \frac{1}{2} \left\| |T|^{2r} + \left[T^2 \left(T^* \right)^2 \right]^r \right\| \tag{3.101}$$

and

$$w^r \left(T |T| T \right) \leq \frac{1}{2} \left\| |T|^{2r} + |T^*|^{4r} \right\|. \tag{3.102}$$

Chapter 4
Trace Inequalities

In this chapter, after recalling some fundamental facts on *Hilbert–Schmidt operators,* *trace operators* and some properties of traces of such operators, we present a trace version of Kato's inequality. Some natural functionals associated to this inequality and their superadditivity and monotonicity are established. Several inequalities for sequences of operators and power series of operators are given as well.

4.1 Trace of Operators

Let $(H, \langle \cdot, \cdot \rangle)$ be a complex Hilbert space and $\{e_i\}_{i \in I}$ an *orthonormal basis* of H. We say that $A \in \mathcal{B}(H)$ is a *Hilbert–Schmidt operator* if

$$\sum_{i \in I} \| Ae_i \|^2 < \infty. \tag{4.1}$$

It is well know that, if $\{e_i\}_{i \in I}$ and $\{f_j\}_{j \in J}$ are orthonormal bases for H and $A \in \mathcal{B}(H)$ then

$$\sum_{i \in I} \| Ae_i \|^2 = \sum_{j \in J} \| Af_j \|^2 = \sum_{j \in J} \| A^* f_j \|^2 \tag{4.2}$$

showing that the definition (4.1) is independent of the orthonormal basis and A is a Hilbert–Schmidt operator iff A^* is a Hilbert–Schmidt operator.

Let $\mathcal{B}_2(H)$ the set of Hilbert–Schmidt operators in $\mathcal{B}(H)$. For $A \in \mathcal{B}_2(H)$ we define

$$\| A \|_2 := \left(\sum_{i \in I} \| Ae_i \|^2 \right)^{1/2} \tag{4.3}$$

for $\{e_i\}_{i \in I}$ an orthonormal basis of H. This definition does not depend on the choice of the orthonormal basis.

© The Author(s), under exclusive license to Springer Nature Switzerland AG 2019
S. S. Dragomir, *Kato's Type Inequalities for Bounded Linear*
Operators in Hilbert Spaces, SpringerBriefs in Mathematics,
https://doi.org/10.1007/978-3-030-17459-0_4

Using the triangle inequality in $l^2\left(I\right)$, one checks that $\mathcal{B}_2\left(H\right)$ is a *vector space* and that $\|\cdot\|_2$ is a norm on $\mathcal{B}_2\left(H\right)$, which is usually called in the literature as the *Hilbert–Schmidt norm*.

Denote *the modulus* of an operator $A \in \mathcal{B}\left(H\right)$ by $|A| := (A^*A)^{1/2}$.

Because $\||A|\,x\| = \|Ax\|$ for all $x \in H$, A is Hilbert–Schmidt iff $|A|$ is Hilbert–Schmidt and $\|A\|_2 = \||A|\|_2$. From (4.2) we have that if $A \in \mathcal{B}_2\left(H\right)$, then $A^* \in \mathcal{B}_2\left(H\right)$ and $\|A\|_2 = \|A^*\|_2$.

The following theorem collects some of the most important properties of Hilbert–Schmidt operators:

Theorem 4.1 *We have*

(i) $(\mathcal{B}_2\left(H\right), \|\cdot\|_2)$ *is a Hilbert space with inner product*

$$\langle A, B \rangle_2 := \sum_{i \in I} \langle Ae_i, Be_i \rangle = \sum_{i \in I} \langle B^*Ae_i, e_i \rangle \tag{4.4}$$

and the definition does not depend on the choice of the orthonormal basis $\{e_i\}_{i \in I}$;

(ii) We have the inequalities

$$\|A\| \le \|A\|_2 \tag{4.5}$$

for any $A \in \mathcal{B}_2\left(H\right)$ *and*

$$\|AT\|_2, \|TA\|_2 \le \|T\|\,\|A\|_2 \tag{4.6}$$

for any $A \in \mathcal{B}_2\left(H\right)$ *and* $T \in \mathcal{B}\left(H\right)$;

(iii) $\mathcal{B}_2\left(H\right)$ *is an operator ideal in* $\mathcal{B}\left(H\right)$, *i.e.*

$$\mathcal{B}\left(H\right)\mathcal{B}_2\left(H\right)\mathcal{B}\left(H\right) \subseteq \mathcal{B}_2\left(H\right);$$

(iv) $\mathcal{B}_{fin}\left(H\right)$, *the space of operators of finite rank, is a dense subspace of* $\mathcal{B}_2\left(H\right)$;

(v) $\mathcal{B}_2\left(H\right) \subseteq \mathcal{K}\left(H\right)$, *where* $\mathcal{K}\left(H\right)$ *denotes the algebra of compact operators on* H.

If $\{e_i\}_{i \in I}$ an orthonormal basis of H, we say that $A \in \mathcal{B}\left(H\right)$ is *trace class* if

$$\|A\|_1 := \sum_{i \in I} \langle |A|\,e_i, e_i \rangle < \infty. \tag{4.7}$$

The definition of $\|A\|_1$ does not depend on the choice of the orthonormal basis $\{e_i\}_{i \in I}$. We denote by $\mathcal{B}_1\left(H\right)$ the set of trace class operators in $\mathcal{B}\left(H\right)$.

The following proposition holds:

Proposition 4.2 *If* $A \in \mathcal{B}\left(H\right)$, *then the following are equivalent:*

(i) $A \in \mathcal{B}_1\left(H\right)$;

(ii) $|A|^{1/2} \in \mathcal{B}_2\left(H\right)$;

(ii) A *(or* $|A|$*) is the product of two elements of* $\mathcal{B}_2\left(H\right)$.

The following properties are also well known:

Theorem 4.3 *With the above notations:*
(i) We have

$$\|A\|_1 = \|A^*\|_1 \ \text{and} \ \|A\|_2 \leq \|A\|_1 \tag{4.8}$$

for any $A \in \mathcal{B}_1(H)$;
(ii) $\mathcal{B}_1(H)$ is an operator ideal in $\mathcal{B}(H)$, i.e.

$$\mathcal{B}(H)\mathcal{B}_1(H)\mathcal{B}(H) \subseteq \mathcal{B}_1(H);$$

(iii) We have

$$\mathcal{B}_2(H)\mathcal{B}_2(H) = \mathcal{B}_1(H);$$

(iv) We have

$$\|A\|_1 = \sup\{|\langle A, B\rangle_2| \mid B \in \mathcal{B}_2(H)|, \ \|B\| \leq 1\};$$

(v) $(\mathcal{B}_1(H), \|\cdot\|_1)$ is a Banach space.
(iv) We have the following isometric isomorphisms

$$\mathcal{B}_1(H) \cong K(H)^* \ \text{and} \ \mathcal{B}_1(H)^* \cong \mathcal{B}(H),$$

where $K(H)^*$ is the dual space of $K(H)$ and $\mathcal{B}_1(H)^*$ is the dual space of $\mathcal{B}_1(H)$.

We define the *trace* of a trace class operator $A \in \mathcal{B}_1(H)$ to be

$$\text{tr}(A) := \sum_{i \in I} \langle Ae_i, e_i\rangle, \tag{4.9}$$

where $\{e_i\}_{i \in I}$ an orthonormal basis of H. Note that this coincides with the usual definition of the trace if H is finite-dimensional. We observe that the series (4.9) converges absolutely and it is independent from the choice of basis.

The following result collects some properties of the trace:

Theorem 4.4 *We have*
(i) If $A \in \mathcal{B}_1(H)$ then $A^* \in \mathcal{B}_1(H)$ and

$$\text{tr}(A^*) = \overline{\text{tr}(A)}; \tag{4.10}$$

(ii) If $A \in \mathcal{B}_1(H)$ and $T \in \mathcal{B}(H)$, then $AT, TA \in \mathcal{B}_1(H)$ and

$$\text{tr}(AT) = \text{tr}(TA) \ \text{and} \ |\text{tr}(AT)| \leq \|A\|_1 \|T\|; \tag{4.11}$$

(iii) $\text{tr}(\cdot)$ is a bounded linear functional on $\mathcal{B}_1(H)$ with $\|\text{tr}\| = 1$;

(iv) If A, B $\in \mathcal{B}_2(H)$ then AB, $BA \in \mathcal{B}_1(H)$ and $\mathrm{tr}(AB) = \mathrm{tr}(BA)$;
(v) $\mathcal{B}_{fin}(H)$ is a dense subspace of $\mathcal{B}_1(H)$.

Utilising the trace notation we obviously have that

$$\langle A, B \rangle_2 = \mathrm{tr}\left(B^*A\right) = \mathrm{tr}\left(AB^*\right) \text{ and } \|A\|_2^2 = \mathrm{tr}\left(A^*A\right) = \mathrm{tr}\left(|A|^2\right)$$

for any A, $B \in \mathcal{B}_2(H)$.

For the theory of trace functionals and their applications the reader is referred to [31].

4.2   Trace Inequalities via Kato's Result

We start with the following result:

Theorem 4.5 (Dragomir [8]) *Let $T \in \mathcal{B}(H)$.*
(i) If for some $\alpha \in (0, 1)$ we have $|T|^{2\alpha}$, $|T^*|^{2(1-\alpha)} \in \mathcal{B}_1(H)$, then $T \in \mathcal{B}_1(H)$ and we have the inequality

$$|\mathrm{tr}(T)|^2 \leq \mathrm{tr}\left(|T|^{2\alpha}\right) \mathrm{tr}\left(\left|T^*\right|^{2(1-\alpha)}\right); \tag{4.12}$$

(ii) If for some $\alpha \in [0, 1]$ and an orthonormal basis $\{e_i\}_{i \in I}$ the sum

$$\sum_{i \in I} \|T e_i\|^\alpha \left\|T^* e_i\right\|^{1-\alpha}$$

is finite, then $T \in \mathcal{B}_1(H)$ and we have the inequality

$$|\mathrm{tr}(T)| \leq \sum_{i \in I} \|T e_i\|^\alpha \left\|T^* e_i\right\|^{1-\alpha}. \tag{4.13}$$

Moreover, if the sums $\sum_{i \in I} \|T e_i\|$ and $\sum_{i \in I} \|T^* e_i\|$ are finite for an orthonormal basis $\{e_i\}_{i \in I}$, then $T \in \mathcal{B}_1(H)$ and we have

$$|\mathrm{tr}(T)| \leq \inf_{\alpha \in [0,1]} \left\{\sum_{i \in I} \|T e_i\|^\alpha \left\|T^* e_i\right\|^{1-\alpha}\right\} \leq \min\left\{\sum_{i \in F} \|T e_i\|, \sum_{i \in F} \|T^* e_i\|\right\}. \tag{4.14}$$

Proof (i) Assume that $\alpha \in (0, 1)$. Let $\{e_i\}_{i \in I}$ be an orthonormal basis in H and F a finite part of I. Then by Kato's inequality (K) we have

$$\left|\sum_{i \in F} \langle T e_i, e_i \rangle\right| \leq \sum_{i \in F} |\langle T e_i, e_i \rangle| \leq \sum_{i \in F} \langle |T|^{2\alpha} e_i, e_i \rangle^{1/2} \left(\left|T^*\right|^{2(1-\alpha)} e_i, e_i\right)^{1/2}. \tag{4.15}$$

By Cauchy–Buniakovski–Schwarz inequality for finite sums we have

$$\sum_{i \in F} \langle |T|^{2\alpha} e_i, e_i \rangle^{1/2} \langle |T^*|^{2(1-\alpha)} e_i, e_i \rangle^{1/2} \tag{4.16}$$

$$\leq \left(\sum_{i \in F} \left[\langle |T|^{2\alpha} e_i, e_i \rangle^{1/2} \right]^2 \right)^{1/2} \left(\sum_{i \in F} \left[\langle |T^*|^{2(1-\alpha)} e_i, e_i \rangle^{1/2} \right]^2 \right)^{1/2}$$

$$= \left(\sum_{i \in F} \langle |T|^{2\alpha} e_i, e_i \rangle \right)^{1/2} \left(\sum_{i \in F} \langle |T^*|^{2(1-\alpha)} e_i, e_i \rangle \right)^{1/2} .$$

Therefore, by (4.15) and (4.16) we have

$$\left| \sum_{i \in F} \langle T e_i, e_i \rangle \right| \leq \left(\sum_{i \in F} \langle |T|^{2\alpha} e_i, e_i \rangle \right)^{1/2} \left(\sum_{i \in F} \langle |T^*|^{2(1-\alpha)} e_i, e_i \rangle \right)^{1/2} \tag{4.17}$$

for any finite part F of I.

If for some $\alpha \in (0, 1)$ we have $|T|^{2\alpha}, |T^*|^{2(1-\alpha)} \in \mathcal{B}_1(H)$, then the sums $\sum_{i \in I} \langle |T|^{2\alpha} e_i, e_i \rangle$ and $\sum_{i \in I} \langle |T^*|^{2(1-\alpha)} e_i, e_i \rangle$ are finite and by (4.17) we have that $\sum_{i \in I} \langle T e_i, e_i \rangle$ is also finite and we have the inequality (4.12).

(ii) Assume that $\alpha \in [0, 1]$. Let $\{e_i\}_{i \in I}$ be an orthonormal basis in H and F a finite part of I. Utilising McCarthy's inequality for the positive operator P, namely

$$\langle P^\beta x, x \rangle \leq \langle Px, x \rangle^\beta ,$$

that holds for $\beta \in [0, 1]$ and $x \in H$, $\|x\| = 1$, we have

$$\langle |T|^{2\alpha} e_i, e_i \rangle \leq \langle |T|^2 e_i, e_i \rangle^\alpha$$

and

$$\langle |T^*|^{2(1-\alpha)} e_i, e_i \rangle \leq \langle |T^*|^2 e_i, e_i \rangle^{1-\alpha}$$

for any $i \in I$.

Making use of (4.15) we have

$$\left| \sum_{i \in F} \langle T e_i, e_i \rangle \right| \leq \sum_{i \in F} |\langle T e_i, e_i \rangle| \leq \sum_{i \in F} \langle |T|^{2\alpha} e_i, e_i \rangle^{1/2} \langle |T^*|^{2(1-\alpha)} e_i, e_i \rangle^{1/2} \tag{4.18}$$

$$\leq \sum_{i \in F} \langle |T|^2 e_i, e_i \rangle^{\alpha/2} \langle |T^*|^2 e_i, e_i \rangle^{(1-\alpha)/2}$$

$$= \sum_{i \in F} \langle T^* T e_i, e_i \rangle^{\alpha/2} \langle T T^* e_i, e_i \rangle^{(1-\alpha)/2}$$

$$= \sum_{i \in F} \|T e_i\|^\alpha \|T^* e_i\|^{1-\alpha} .$$

Utilizing Hölder's inequality for finite sums and $p = \frac{1}{\alpha}$, $q = \frac{1}{1-\alpha}$ we also have

$$\sum_{i \in F} \|T e_i\|^\alpha \|T^* e_i\|^{1-\alpha} \tag{4.19}$$

$$\leq \left[\sum_{i \in F} (\|T e_i\|^\alpha)^{1/\alpha} \right]^\alpha \left[\sum_{i \in F} \left(\|T^* e_i\|^{1-\alpha} \right)^{1/(1-\alpha)} \right]^{1-\alpha}$$

$$= \left[\sum_{i \in F} \|T e_i\| \right]^\alpha \left[\sum_{i \in F} \|T^* e_i\| \right]^{1-\alpha}.$$

Since all the series involved in (4.18) and (4.19) are convergent, then we get

$$\left| \sum_{i \in I} \langle T e_i, e_i \rangle \right| \leq \sum_{i \in I} \|T e_i\|^\alpha \|T^* e_i\|^{1-\alpha} \tag{4.20}$$

$$\leq \left[\sum_{i \in I} \|T e_i\| \right]^\alpha \left[\sum_{i \in I} \|T^* e_i\| \right]^{1-\alpha}$$

for any $\alpha \in [0, 1]$.

Taking the infimum over $\alpha \in [0, 1]$ in (4.20) produces

$$\left| \sum_{i \in I} \langle T e_i, e_i \rangle \right| \leq \inf_{\alpha \in [0,1]} \left\{ \sum_{i \in F} \|T e_i\|^\alpha \|T^* e_i\|^{1-\alpha} \right\} \tag{4.21}$$

$$\leq \inf_{\alpha \in [0,1]} \left[\sum_{i \in F} \|T e_i\| \right]^\alpha \left[\sum_{i \in F} \|T^* e_i\| \right]^{1-\alpha}$$

$$= \min \left\{ \sum_{i \in F} \|T e_i\|, \sum_{i \in F} \|T^* e_i\| \right\}.$$

∎

Corollary 4.6 (Dragomir [8]) *Let $T \in \mathcal{B}(H)$.*

(i) If we have $|T|, |T^*| \in \mathcal{B}_1(H)$, then $T \in \mathcal{B}_1(H)$ and we have the inequality

$$|\operatorname{tr}(T)|^2 \leq \operatorname{tr}(|T|) \operatorname{tr}(|T^*|); \tag{4.22}$$

(ii) If for an orthonormal basis $\{e_i\}_{i \in I}$ the sum $\sum_{i \in I} \sqrt{\|T e_i\| \|T^* e_i\|}$ is finite, then $T \in \mathcal{B}_1(H)$ and we have the inequality

$$|\operatorname{tr}(T)| \leq \sum_{i \in I} \sqrt{\|T e_i\| \|T^* e_i\|}. \tag{4.23}$$

Corollary 4.7 (Dragomir [8]) *Let* $N \in \mathcal{B}(H)$ *be a normal operator. If for some* $\alpha \in (0, 1)$ *we have* $|N|^{2\alpha}$, $|N|^{2(1-\alpha)} \in \mathcal{B}_1(H)$, *then* $N \in \mathcal{B}_1(H)$ *and we have the inequality*

$$|\operatorname{tr}(N)|^2 \leq \operatorname{tr}\left(|N|^{2\alpha}\right) \operatorname{tr}\left(|N|^{2(1-\alpha)}\right). \tag{4.24}$$

In particular, if $|N| \in \mathcal{B}_1(H)$, *then* $N \in \mathcal{B}_1(H)$ *and*

$$|\operatorname{tr}(N)| \leq \operatorname{tr}(|N|). \tag{4.25}$$

The following result also holds.

Theorem 4.8 (Dragomir [8]) *Let* $T \in \mathcal{B}(H)$ *and* $A, B \in \mathcal{B}_2(H)$.
  (i) For any $\alpha \in [0, 1]$ *we have* $|A^*|^2 |T|^{2\alpha}$, $|B^*|^2 |T^*|^{2(1-\alpha)}$ *and* $B^*TA \in \mathcal{B}_1(H)$
and

$$\left|\operatorname{tr}\left(AB^*T\right)\right|^2 \leq \operatorname{tr}\left(|A^*|^2 |T|^{2\alpha}\right) \operatorname{tr}\left(|B^*|^2 |T^*|^{2(1-\alpha)}\right); \tag{4.26}$$

(ii) We also have

$$\left|\operatorname{tr}\left(AB^*T\right)\right|^2 \tag{4.27}$$
$$\leq \min\left\{\operatorname{tr}\left(|B|^2\right) \operatorname{tr}\left(|A^*|^2 |T|^2\right), \operatorname{tr}\left(|A|^2\right) \operatorname{tr}\left(|B^*|^2 |T^*|^2\right)\right\}.$$

Proof (i) Let $\{e_i\}_{i \in I}$ be an orthonormal basis in H and F a finite part of I. Then by Kato's inequality (K) we have

$$|\langle TAe_i, Be_i\rangle|^2 \leq \langle |T|^{2\alpha} Ae_i, Ae_i\rangle \left\langle |T^*|^{2(1-\alpha)} Be_i, Be_i\right\rangle \tag{4.28}$$

for any $i \in I$. This is equivalent to

$$\left|\langle B^*TAe_i, e_i\rangle\right| \leq \langle A^* |T|^{2\alpha} Ae_i, e_i\rangle^{1/2} \left\langle B^* |T^*|^{2(1-\alpha)} Be_i, e_i\right\rangle^{1/2} \tag{4.29}$$

for any $i \in I$.
  Using the generalized triangle inequality for the modulus and the Cauchy–Bunyakowsky–Schwarz inequality for finite sums we have from (4.29) that

$$\left|\sum_{i \in F}\langle B^*TAe_i, e_i\rangle\right| \tag{4.30}$$
$$\leq \sum_{i \in F} |\langle B^*TAe_i, e_i\rangle|$$
$$\leq \sum_{i \in F} \langle A^* |T|^{2\alpha} Ae_i, e_i\rangle^{1/2} \left\langle B^* |T^*|^{2(1-\alpha)} Be_i, e_i\right\rangle^{1/2}$$
$$\leq \left[\sum_{i \in F}\left(\langle A^* |T|^{2\alpha} Ae_i, e_i\rangle^{1/2}\right)^2\right]^{1/2}$$

$$\times \left[\sum_{i \in F} \left(\left\langle B^* |T^*|^{2(1-\alpha)} Be_i, e_i \right\rangle^{1/2} \right)^2 \right]^{1/2}$$

$$= \left[\sum_{i \in F} \left\langle A^* |T|^{2\alpha} Ae_i, e_i \right\rangle \right]^{1/2} \left[\sum_{i \in F} \left\langle B^* |T^*|^{2(1-\alpha)} Be_i, e_i \right\rangle \right]^{1/2}$$

for any F a finite part of I.

Let $\alpha \in [0, 1]$. Since $A, B \in \mathcal{B}_2(H)$, then $A^* |T|^{2\alpha} A$, $B^* |T^*|^{2(1-\alpha)} B$ and $B^* T A \in \mathcal{B}_1(H)$ and by (4.30) we have

$$\left| \mathrm{tr}\left(B^* T A \right) \right| \leq \left[\mathrm{tr}\left(A^* |T|^{2\alpha} A \right) \right]^{1/2} \left[\mathrm{tr}\left(B^* |T^*|^{2(1-\alpha)} B \right) \right]^{1/2}. \qquad (4.31)$$

Since, by the properties of trace we have

$$\mathrm{tr}\left(B^* T A \right) = \mathrm{tr}\left(A B^* T \right),$$

$$\mathrm{tr}\left(A^* |T|^{2\alpha} A \right) = \mathrm{tr}\left(A A^* |T|^{2\alpha} \right) = \mathrm{tr}\left(|A^*|^2 |T|^{2\alpha} \right)$$

and

$$\mathrm{tr}\left(B^* |T^*|^{2(1-\alpha)} B \right) = \mathrm{tr}\left(|B^*|^2 |T^*|^{2(1-\alpha)} \right),$$

then by (4.31) we get (4.26).

(ii) Utilising McCarthy's inequality [29] for the positive operator P

$$\left\langle P^\beta x, x \right\rangle \leq \left\langle Px, x \right\rangle^\beta$$

that holds for $\beta \in (0, 1)$ and $x \in H$, $\|x\| = 1$, we have

$$\left\langle P^\beta y, y \right\rangle \leq \|y\|^{2(1-\beta)} \left\langle Py, y \right\rangle^\beta \qquad (4.32)$$

for any $y \in H$.

Let $\{e_i\}_{i \in I}$ be an orthonormal basis in H and F a finite part of I. From (4.32) we have

$$\left\langle |T|^{2\alpha} Ae_i, Ae_i \right\rangle \leq \|Ae_i\|^{2(1-\alpha)} \left\langle |T|^2 Ae_i, Ae_i \right\rangle^\alpha$$

and

$$\left\langle |T^*|^{2(1-\alpha)} Be_i, Be_i \right\rangle \leq \|Be_i\|^{2\alpha} \left\langle |T^*|^2 Be_i, Be_i \right\rangle^{1-\alpha}$$

for any $i \in I$.

Making use of the inequality (4.28) we get

$$|\langle T Ae_i, Be_i \rangle|^2 \leq \|Ae_i\|^{2(1-\alpha)} \left\langle |T|^2 Ae_i, Ae_i \right\rangle^\alpha \|Be_i\|^{2\alpha} \left\langle |T^*|^2 Be_i, Be_i \right\rangle^{1-\alpha}$$

$$= \|Be_i\|^{2\alpha} \langle |T|^2 Ae_i, Ae_i \rangle^{\alpha} \|Ae_i\|^{2(1-\alpha)} \langle |T^*|^2 Be_i, Be_i \rangle^{1-\alpha}$$

and taking the square root we get

$$|\langle TAe_i, Be_i \rangle| \le \|Be_i\|^{\alpha} \langle |T|^2 Ae_i, Ae_i \rangle^{\frac{\alpha}{2}} \|Ae_i\|^{1-\alpha} \langle |T^*|^2 Be_i, Be_i \rangle^{\frac{1-\alpha}{2}} \quad (4.33)$$

for any $i \in I$.

Using the generalized triangle inequality for the modulus and the Hölder's inequality for finite sums and $p = \frac{1}{\alpha}$, $q = \frac{1}{1-\alpha}$ we get from (4.33) that

$$\left| \sum_{i \in F} \langle B^* T Ae_i, e_i \rangle \right| \tag{4.34}$$

$$\le \sum_{i \in F} |\langle B^* T Ae_i, e_i \rangle|$$

$$\le \sum_{i \in F} \|Be_i\|^{\alpha} \langle |T|^2 Ae_i, Ae_i \rangle^{\frac{\alpha}{2}} \|Ae_i\|^{1-\alpha} \langle |T^*|^2 Be_i, Be_i \rangle^{\frac{1-\alpha}{2}}$$

$$\le \left(\sum_{i \in F} \left[\|Be_i\|^{\alpha} \langle |T|^2 Ae_i, Ae_i \rangle^{\frac{\alpha}{2}} \right]^{1/\alpha} \right)^{\alpha}$$

$$\times \left(\sum_{i \in F} \left[\|Ae_i\|^{1-\alpha} \langle |T^*|^2 Be_i, Be_i \rangle^{\frac{1-\alpha}{2}} \right]^{1/(1-\alpha)} \right)^{1-\alpha}$$

$$= \left(\sum_{i \in F} \|Be_i\| \langle |T|^2 Ae_i, Ae_i \rangle^{\frac{1}{2}} \right)^{\alpha} \left(\sum_{i \in F} \|Ae_i\| \langle |T^*|^2 Be_i, Be_i \rangle^{\frac{1}{2}} \right)^{1-\alpha}.$$

By Cauchy–Bunyakowsky–Schwarz inequality for finite sums we also have

$$\sum_{i \in F} \|Be_i\| \langle |T|^2 Ae_i, Ae_i \rangle^{\frac{1}{2}} \le \left(\sum_{i \in F} \|Be_i\|^2 \right)^{1/2} \left(\sum_{i \in F} \langle |T|^2 Ae_i, Ae_i \rangle \right)^{1/2}$$

$$= \left(\sum_{i \in F} \langle |B|^2 e_i, e_i \rangle \right)^{1/2} \left(\sum_{i \in F} \langle A^* |T|^2 Ae_i, e_i \rangle \right)^{1/2}$$

and

$$\sum_{i \in F} \|Ae_i\| \langle |T^*|^2 Be_i, Be_i \rangle^{\frac{1}{2}} \le \left(\sum_{i \in F} \|Ae_i\|^2 \right)^{1/2} \left(\sum_{i \in F} \langle |T^*|^2 Be_i, Be_i \rangle \right)^{1/2}$$

$$= \left(\sum_{i \in F} \langle |A|^2 e_i, e_i \rangle \right)^{1/2} \left(\sum_{i \in F} \langle B^* |T^*|^2 Be_i, e_i \rangle \right)^{1/2}$$

and by (4.34) we obtain

$$\left| \sum_{i \in F} \langle B^* T A e_i, e_i \rangle \right| \tag{4.35}$$

$$\leq \left(\sum_{i \in F} \langle |B|^2 e_i, e_i \rangle \right)^{\alpha/2} \left(\sum_{i \in F} \langle A^* |T|^2 A e_i, e_i \rangle \right)^{\alpha/2}$$

$$\times \left(\sum_{i \in F} \langle |A|^2 e_i, e_i \rangle \right)^{(1-\alpha)/2} \left(\sum_{i \in F} \langle B^* |T^*|^2 B e_i, e_i \rangle \right)^{(1-\alpha)/2}$$

for any F a finite part of I.

Let $\alpha \in [0, 1]$. Since $A, B \in \mathcal{B}_2(H)$, then $A^* |T|^2 A$ and $B^* |T^*|^2 B \in \mathcal{B}_1(H)$ and by (4.35) we get

$$\left| \operatorname{tr} \left(A B^* T \right) \right|^2 \tag{4.36}$$

$$\leq \left[\operatorname{tr} \left(|B|^2 \right) \operatorname{tr} \left(A^* |T|^2 A \right) \right]^{\alpha} \left[\operatorname{tr} \left(|A|^2 \right) \operatorname{tr} \left(B^* |T^*|^2 B \right) \right]^{1-\alpha}$$

$$= \left[\operatorname{tr} \left(|B|^2 \right) \operatorname{tr} \left(|A^*|^2 |T|^2 \right) \right]^{\alpha} \left[\operatorname{tr} \left(|A|^2 \right) \operatorname{tr} \left(|B^*|^2 |T^*|^2 \right) \right]^{1-\alpha}.$$

Taking the infimum over $\alpha \in [0, 1]$ we get (4.27).   ∎

Corollary 4.9 (Dragomir [8]) *Let $T \in \mathcal{B}(H)$ and $A, B \in \mathcal{B}_2(H)$. We have $|A^*|^2 |T|, |B^*|^2 |T^*|$ and $B^* T A \in \mathcal{B}_1(H)$ and*

$$\left| \operatorname{tr} \left(A B^* T \right) \right|^2 \leq \operatorname{tr} \left(|A^*|^2 |T| \right) \operatorname{tr} \left(|B^*|^2 |T^*| \right). \tag{4.37}$$

Corollary 4.10 (Dragomir [8]) *Let $N \in \mathcal{B}(H)$ be a normal operator and $A, B \in \mathcal{B}_2(H)$.*

(i) For any $\alpha \in [0, 1]$ we have $|A^*|^2 |N|^{2\alpha}, |B^*|^2 |N|^{2(1-\alpha)}$ and $B^* N A \in \mathcal{B}_1(H)$ and

$$\left| \operatorname{tr} \left(A B^* N \right) \right|^2 \leq \operatorname{tr} \left(|A^*|^2 |N|^{2\alpha} \right) \operatorname{tr} \left(|B^*|^2 |N|^{2(1-\alpha)} \right). \tag{4.38}$$

In particular, we have $|A^*|^2 |N|, |B^*|^2 |N|$ and $B^* N A \in \mathcal{B}_1(H)$ and

$$\left| \operatorname{tr} \left(A B^* N \right) \right|^2 \leq \operatorname{tr} \left(|A^*|^2 |N| \right) \operatorname{tr} \left(|B^*|^2 |N| \right). \tag{4.39}$$

(ii) We also have

$$\left| \operatorname{tr} \left(A B^* N \right) \right|^2 \tag{4.40}$$

$$\leq \min \left\{ \operatorname{tr} \left(|B|^2 \right) \operatorname{tr} \left(|A^*|^2 |N|^2 \right), \operatorname{tr} \left(|A|^2 \right) \operatorname{tr} \left(|B^*|^2 |N|^2 \right) \right\}.$$

Remark 4.11 Let $\alpha \in [0, 1]$. By replacing A with A^* and B with B^* in (4.26) we get

$$\left| \text{tr} \left(A^* B T \right) \right|^2 \le \text{tr} \left(|A|^2 |T|^{2\alpha} \right) \text{tr} \left(|B|^2 |T^*|^{2(1-\alpha)} \right) \qquad (4.41)$$

for any $T \in \mathcal{B}(H)$ *and* $A, B \in \mathcal{B}_2(H)$.

If in this inequality we take $A = B$, then we get

$$\left| \text{tr} \left(|B|^2 T \right) \right|^2 \le \text{tr} \left(|B|^2 |T|^{2\alpha} \right) \text{tr} \left(|B|^2 |T^*|^{2(1-\alpha)} \right) \qquad (4.42)$$

for any $T \in \mathcal{B}(H)$ *and* $B \in \mathcal{B}_2(H)$.

If in (4.41) we take $A = B^*$, then we get

$$\left| \text{tr} \left(B^2 T \right) \right|^2 \le \text{tr} \left(|B^*|^2 |T|^{2\alpha} \right) \text{tr} \left(|B|^2 |T^*|^{2(1-\alpha)} \right) \qquad (4.43)$$

for any $T \in \mathcal{B}(H)$ *and* $B \in \mathcal{B}_2(H)$.

Also, if $T = N$, a normal operator, then (4.42) and (4.43) become

$$\left| \text{tr} \left(|B|^2 N \right) \right|^2 \le \text{tr} \left(|B|^2 |N|^{2\alpha} \right) \text{tr} \left(|B|^2 |N|^{2(1-\alpha)} \right) \qquad (4.44)$$

and

$$\left| \text{tr} \left(B^2 N \right) \right|^2 \le \text{tr} \left(|B^*|^2 |N|^{2\alpha} \right) \text{tr} \left(|B|^2 |N|^{2(1-\alpha)} \right), \qquad (4.45)$$

for any $B \in \mathcal{B}_2(H)$.

4.3 Some Functional Properties

Let $A \in \mathcal{B}_2(H)$ and $P \in \mathcal{B}(H)$ with $P \ge 0$. Then $Q := A^* P A \in \mathcal{B}_1(H)$ with $Q \ge 0$ and writing the inequality (4.42) for $B = (A^* P A)^{1/2} \in \mathcal{B}_2(H)$ we get

$$\left| \text{tr} \left(A^* P A T \right) \right|^2 \le \text{tr} \left(A^* P A |T|^{2\alpha} \right) \text{tr} \left(A^* P A |T^*|^{2(1-\alpha)} \right),$$

which, by the properties of trace, is equivalent to

$$\left| \text{tr} \left(P A T A^* \right) \right|^2 \le \text{tr} \left(P A |T|^{2\alpha} A^* \right) \text{tr} \left(P A |T^*|^{2(1-\alpha)} A^* \right), \qquad (4.46)$$

where $T \in \mathcal{B}(H)$ and $\alpha \in [0, 1]$.

For a given $A \in \mathcal{B}_2(H)$, $T \in \mathcal{B}(H)$ and $\alpha \in [0, 1]$, we consider the functional $\sigma_{A,T,\alpha}$ defined on the cone $\mathcal{B}_+(H)$ of nonnegative operators on $\mathcal{B}(H)$ by

$$\sigma_{A,T,\alpha}(P) := \left[\text{tr} \left(P A |T|^{2\alpha} A^* \right) \right]^{1/2} \left[\text{tr} \left(P A |T^*|^{2(1-\alpha)} A^* \right) \right]^{1/2}$$
$$- \left| \text{tr} \left(P A T A^* \right) \right|.$$

The following theorem collects some fundamental properties of this functional.

Theorem 4.12 (Dragomir [8]) *Let* $A \in \mathcal{B}_2(H)$, $T \in \mathcal{B}(H)$ *and* $\alpha \in [0, 1]$.
(i) *For any* $P, Q \in \mathcal{B}_+(H)$ *we have*

$$\sigma_{A,T,\alpha}(P + Q) \geq \sigma_{A,T,\alpha}(P) + \sigma_{A,T,\alpha}(Q) (\geq 0), \tag{4.47}$$

namely, $\sigma_{A,T,\alpha}$ *is a superadditive functional on* $\mathcal{B}_+(H)$;
(ii) *For any* $P, Q \in \mathcal{B}_+(H)$ *with* $P \geq Q$ *we have*

$$\sigma_{A,T,\alpha}(P) \geq \sigma_{A,T,\alpha}(Q) (\geq 0), \tag{4.48}$$

namely, $\sigma_{A,T,\alpha}$ *is a monotonic nondecreasing functional on* $\mathcal{B}_+(H)$;
(iii) *If* $P, Q \in \mathcal{B}_+(H)$ *and there exist the constants* $M > m > 0$ *such that* $MQ \geq P \geq mQ$ *then*

$$M\sigma_{A,T,\alpha}(Q) \geq \sigma_{A,T,\alpha}(P) \geq m\sigma_{A,T,\alpha}(Q) (\geq 0). \tag{4.49}$$

Proof (i) Let $P, Q \in \mathcal{B}_+(H)$. On utilizing the elementary inequality

$$\left(a^2 + b^2\right)^{1/2} \left(c^2 + d^2\right)^{1/2} \geq ac + bd, \ a, b, c, d \geq 0$$

and the triangle inequality for the modulus, we have

$$
\begin{aligned}
&\sigma_{A,T,\alpha}(P + Q) \\
&= \left[\operatorname{tr}\left((P + Q) A |T|^{2\alpha} A^*\right)\right]^{1/2} \left[\operatorname{tr}\left((P + Q) A |T^*|^{2(1-\alpha)} A^*\right)\right]^{1/2} \\
&\quad - \left|\operatorname{tr}\left((P + Q) A T A^*\right)\right| \\
&= \left[\operatorname{tr}\left(P A |T|^{2\alpha} A^* + Q A |T|^{2\alpha} A^*\right)\right]^{1/2} \\
&\quad \times \left[\operatorname{tr}\left(P A |T^*|^{2(1-\alpha)} A^* + Q A |T^*|^{2(1-\alpha)} A^*\right)\right]^{1/2} \\
&\quad - \left|\operatorname{tr}\left(P A T A^* + Q A T A^*\right)\right| \\
&= \left[\operatorname{tr}\left(P A |T|^{2\alpha} A^*\right) + \operatorname{tr}\left(Q A |T|^{2\alpha} A^*\right)\right]^{1/2} \\
&\quad \times \left[\operatorname{tr}\left(P A |T^*|^{2(1-\alpha)} A^*\right) + \operatorname{tr}\left(Q A |T^*|^{2(1-\alpha)} A^*\right)\right]^{1/2} \\
&\quad - \left|\operatorname{tr}\left(P A T A^*\right) + \operatorname{tr}\left(Q A T A^*\right)\right| \\
&\geq \left[\operatorname{tr}\left(P A |T|^{2\alpha} A^*\right)\right]^{1/2} \left[\operatorname{tr}\left(P A |T^*|^{2(1-\alpha)} A^*\right)\right]^{1/2} \\
&\quad + \left[\operatorname{tr}\left(Q A |T|^{2\alpha} A^*\right)\right]^{1/2} \left[\operatorname{tr}\left(Q A |T^*|^{2(1-\alpha)} A^*\right)\right]^{1/2} \\
&\quad - \left|\operatorname{tr}\left(P A T A^*\right)\right| - \left|\operatorname{tr}\left(Q A T A^*\right)\right| \\
&= \sigma_{A,T,\alpha}(P) + \sigma_{A,T,\alpha}(Q)
\end{aligned}
$$

and the inequality (4.47) is proved.

(ii) Let P, $Q \in \mathcal{B}_+ (H)$ with $P \geq Q$. Utilising the superadditivity property we have

$$\sigma_{A,T,\alpha} (P) = \sigma_{A,T,\alpha} ((P - Q) + Q) \geq \sigma_{A,T,\alpha} (P - Q) + \sigma_{A,T,\alpha} (Q)$$
$$\geq \sigma_{A,T,\alpha} (Q)$$

and the inequality (4.48) is obtained.

(iii) From the monotonicity property we have

$$\sigma_{A,T,\alpha} (P) \geq \sigma_{A,T,\alpha} (mQ) = m\sigma_{A,T,\alpha} (Q)$$

and a similar inequality for M, which prove the desired result (4.49). ∎

Corollary 4.13 (Dragomir [8]) *Let* $A \in \mathcal{B}_2 (H)$, $T \in \mathcal{B} (H)$ *and* $\alpha \in [0, 1]$. *If* $P \in \mathcal{B} (H)$ *is such that there exist the constants* $M > m > 0$ *with* $M1_H \geq P \geq m1_H$, *then we have*

$$M \left(\left[\mathrm{tr} \left(A \, |T|^{2\alpha} \, A^* \right) \right]^{1/2} \left[\mathrm{tr} \left(A \, |T^*|^{2(1-\alpha)} \, A^* \right) \right]^{1/2} - \left| \mathrm{tr} \left(ATA^* \right) \right| \right) \quad (4.50)$$
$$\geq \left[\mathrm{tr} \left(PA \, |T|^{2\alpha} \, A^* \right) \right]^{1/2} \left[\mathrm{tr} \left(PA \, |T^*|^{2(1-\alpha)} \, A^* \right) \right]^{1/2} - \left| \mathrm{tr} \left(PATA^* \right) \right|$$
$$\geq m \left(\left[\mathrm{tr} \left(A \, |T|^{2\alpha} \, A^* \right) \right]^{1/2} \left[\mathrm{tr} \left(A \, |T^*|^{2(1-\alpha)} \, A^* \right) \right]^{1/2} - \left| \mathrm{tr} \left(ATA^* \right) \right| \right).$$

For a given $A \in \mathcal{B}_2 (H)$, $T \in \mathcal{B} (H)$ and $\alpha \in [0, 1]$, if we take $P = |V|^2$ with $V \in \mathcal{B} (H)$, we have

$$
\begin{aligned}
\sigma_{A,T,\alpha} \left(|V|^2 \right) &= \left[\mathrm{tr} \left(|V|^2 A \, |T|^{2\alpha} \, A^* \right) \right]^{1/2} \left[\mathrm{tr} \left(|V|^2 A \, |T^*|^{2(1-\alpha)} \, A^* \right) \right]^{1/2} \\
&\quad - \left| \mathrm{tr} \left(|V|^2 ATA^* \right) \right| \\
&= \left[\mathrm{tr} \left(V^*VA \, |T|^{2\alpha} \, A^* \right) \right]^{1/2} \left[\mathrm{tr} \left(V^*VA \, |T^*|^{2(1-\alpha)} \, A^* \right) \right]^{1/2} \\
&\quad - \left| \mathrm{tr} \left(V^*VATA^* \right) \right| \\
&= \left[\mathrm{tr} \left(A^*V^*VA \, |T|^{2\alpha} \right) \right]^{1/2} \left[\mathrm{tr} \left(A^*V^*VA \, |T^*|^{2(1-\alpha)} \right) \right]^{1/2} \\
&\quad - \left| \mathrm{tr} \left(A^*V^*VAT \right) \right| \\
&= \left[\mathrm{tr} \left((VA)^* \, VA \, |T|^{2\alpha} \right) \right]^{1/2} \left[\mathrm{tr} \left((VA)^* \, VA \, |T^*|^{2(1-\alpha)} \right) \right]^{1/2} \\
&\quad - \left| \mathrm{tr} \left((VA)^* \, VAT \right) \right| \\
&= \left[\mathrm{tr} \left(|VA|^2 \, |T|^{2\alpha} \right) \right]^{1/2} \left[\mathrm{tr} \left(|VA|^2 \, |T^*|^{2(1-\alpha)} \right) \right]^{1/2} - \left| \mathrm{tr} \left(|VA|^2 T \right) \right|.
\end{aligned}
$$

Assume that $A \in \mathcal{B}_2 (H)$, $T \in \mathcal{B} (H)$ and $\alpha \in [0, 1]$.

If we use the superadditivity property of the functional $\sigma_{A,T,\alpha}$ we have for any V, $U \in \mathcal{B}(H)$ that

$$\left[\mathrm{tr}\left(\left(|VA|^2 + |UA|^2\right)|T|^{2\alpha}\right)\right]^{1/2}\left[\mathrm{tr}\left(\left(|VA|^2 + |UA|^2\right)|T^*|^{2(1-\alpha)}\right)\right]^{1/2} \quad (4.51)$$
$$- \left|\mathrm{tr}\left(\left(|VA|^2 + |UA|^2\right)T\right)\right|$$
$$\geq \left[\mathrm{tr}\left(|VA|^2|T|^{2\alpha}\right)\right]^{1/2}\left[\mathrm{tr}\left(|VA|^2|T^*|^{2(1-\alpha)}\right)\right]^{1/2} - \left|\mathrm{tr}\left(|VA|^2 T\right)\right|$$
$$+ \left[\mathrm{tr}\left(|UA|^2|T|^{2\alpha}\right)\right]^{1/2}\left[\mathrm{tr}\left(|UA|^2|T^*|^{2(1-\alpha)}\right)\right]^{1/2} - \left|\mathrm{tr}\left(|UA|^2 T\right)\right| \, (\geq 0).$$

Also, if $|V|^2 \geq |U|^2$ with $V, U \in \mathcal{B}(H)$, then

$$\left[\mathrm{tr}\left(|VA|^2|T|^{2\alpha}\right)\right]^{1/2}\left[\mathrm{tr}\left(|VA|^2|T^*|^{2(1-\alpha)}\right)\right]^{1/2} - \left|\mathrm{tr}\left(|VA|^2 T\right)\right| \quad (4.52)$$
$$\geq \left[\mathrm{tr}\left(|UA|^2|T|^{2\alpha}\right)\right]^{1/2}\left[\mathrm{tr}\left(|UA|^2|T^*|^{2(1-\alpha)}\right)\right]^{1/2} - \left|\mathrm{tr}\left(|UA|^2 T\right)\right| \, (\geq 0).$$

If $U \in \mathcal{B}(H)$ is invertible, then

$$\frac{1}{\|U^{-1}\|}\|x\| \leq \|Ux\| \leq \|U\|\,\|x\| \text{ for any } x \in H,$$

which implies that

$$\frac{1}{\|U^{-1}\|^2} 1_H \leq |U|^2 \leq \|U\|^2 1_H.$$

Utilising (4.50) we get

$$\|U\|^2 \left(\left[\mathrm{tr}\left(|A|^2|T|^{2\alpha}\right)\right]^{1/2}\left[\mathrm{tr}\left(|A|^2|T^*|^{2(1-\alpha)}\right)\right]^{1/2} - \left|\mathrm{tr}\left(|A|^2 T\right)\right|\right) \quad (4.53)$$
$$\geq \left[\mathrm{tr}\left(|UA|^2|T|^{2\alpha}\right)\right]^{1/2}\left[\mathrm{tr}\left(|UA|^2|T^*|^{2(1-\alpha)}\right)\right]^{1/2} - \left|\mathrm{tr}\left(|UA|^2 T\right)\right|$$
$$\geq \frac{1}{\|U^{-1}\|^2}\left(\left[\mathrm{tr}\left(|A|^2|T|^{2\alpha}\right)\right]^{1/2}\left[\mathrm{tr}\left(|A|^2|T^*|^{2(1-\alpha)}\right)\right]^{1/2} - \left|\mathrm{tr}\left(|A|^2 T\right)\right|\right).$$

4.4 Inequalities for Sequences of Operators

For $n \geq 2$, define the Cartesian products $\mathcal{B}^{(n)}(H) := \mathcal{B}(H) \times ... \times \mathcal{B}(H)$, $\mathcal{B}_2^{(n)}(H) := \mathcal{B}_2(H) \times ... \times \mathcal{B}_2(H)$ and $\mathcal{B}_+^{(n)}(H) := \mathcal{B}_+(H) \times ... \times \mathcal{B}_+(H)$ where $\mathcal{B}_+(H)$ denotes the convex cone of nonnegative selfadjoint operators on H, i.e. $P \in \mathcal{B}_+(H)$ if $\langle Px, x \rangle \geq 0$ for any $x \in H$.

Proposition 4.14 (Dragomir [8]) *Let* $\mathbf{P} = (P_1, ..., P_n) \in \mathcal{B}_+^{(n)}(H)$, $\mathbf{T} = (T_1, ..., T_n) \in \mathcal{B}^{(n)}(H)$, $\mathbf{A} = (A_1, ..., A_n) \in \mathcal{B}_2^{(n)}(H)$ *and* $\mathbf{z} = (z_1, ..., z_n) \in \mathbb{C}^n$ *with* $n \geq 2$. *Then*

$$\left| \operatorname{tr} \left(\sum_{k=1}^{n} z_k P_k A_k T_k A_k^* \right) \right|^2 \tag{4.54}$$

$$\leq \operatorname{tr} \left(\sum_{k=1}^{n} |z_k| \, P_k A_k \, |T_k|^{2\alpha} \, A_k^* \right) \operatorname{tr} \left(\sum_{k=1}^{n} |z_k| \, P_k A_k \, |T_k^*|^{2(1-\alpha)} \, A_k^* \right)$$

for any $\alpha \in [0, 1]$.

Proof Using the properties of modulus and the inequality (4.46) we have

$$\left| \operatorname{tr} \left(\sum_{k=1}^{n} z_k P_k A_k T_k A_k^* \right) \right|$$

$$= \left| \sum_{k=1}^{n} z_k \operatorname{tr} \left(P_k A_k T_k A_k^* \right) \right| \leq \sum_{k=1}^{n} |z_k| \, \left| \operatorname{tr} \left(P_k A_k T_k A_k^* \right) \right|$$

$$\leq \sum_{k=1}^{n} |z_k| \left[\operatorname{tr} \left(P_k A_k \, |T_k|^{2\alpha} \, A_k^* \right) \right]^{1/2} \left[\operatorname{tr} \left(P_k A_k \, |T_k^*|^{2(1-\alpha)} \, A_k^* \right) \right]^{1/2}.$$

Utilizing the weighted discrete Cauchy–Bunyakovsky–Schwarz inequality we also have

$$\sum_{k=1}^{n} |z_k| \left[\operatorname{tr} \left(P_k A_k \, |T_k|^{2\alpha} \, A_k^* \right) \right]^{1/2} \left[\operatorname{tr} \left(P_k A_k \, |T_k^*|^{2(1-\alpha)} \, A_k^* \right) \right]^{1/2}$$

$$\leq \left(\sum_{k=1}^{n} |z_k| \left(\left[\operatorname{tr} \left(P_k A_k \, |T_k|^{2\alpha} \, A_k^* \right) \right]^{1/2} \right)^2 \right)^{1/2}$$

$$\times \left(\sum_{k=1}^{n} |z_k| \left(\left[\operatorname{tr} \left(P_k A_k \, |T_k^*|^{2(1-\alpha)} \, A_k^* \right) \right]^{1/2} \right)^2 \right)^{1/2}$$

$$= \left(\sum_{k=1}^{n} |z_k| \operatorname{tr} \left(P_k A_k \, |T_k|^{2\alpha} \, A_k^* \right) \right)^{1/2} \left(\sum_{k=1}^{n} |z_k| \operatorname{tr} \left(P_k A_k \, |T_k^*|^{2(1-\alpha)} \, A_k^* \right) \right)^{1/2},$$

which imply the desired result (4.54). ∎

Remark 4.15 If we take $P_k = 1_H$ for any $k \in \{1, ..., n\}$ in (4.54), then we have the simpler inequality

$$\left| \operatorname{tr} \left(\sum_{k=1}^{n} z_k \, |A_k|^2 \, T_k \right) \right|^2 \tag{4.55}$$

$$\leq \operatorname{tr} \left(\sum_{k=1}^{n} |z_k| \, |A_k|^2 \, |T_k|^{2\alpha} \right) \operatorname{tr} \left(\sum_{k=1}^{n} |z_k| \, |A_k|^2 \, |T_k^*|^{2(1-\alpha)} \right)$$

provided that $\mathbf{T} = (T_1, ..., T_n) \in \mathcal{B}^{(n)}(H)$, $\mathbf{A} = (A_1, ..., A_n) \in \mathcal{B}_2^{(n)}(H)$, $\alpha \in [0, 1]$ and $\mathbf{z} = (z_1, ..., z_n) \in \mathbb{C}^n$.

We consider the functional for n-tuples of nonnegative operators $\mathbf{P} = (P_1, ..., P_n) \in \mathcal{B}_+^{(n)}(H)$ as follows:

$$\sigma_{\mathbf{A},\mathbf{T},\alpha}(\mathbf{P}) := \left[\operatorname{tr} \left(\sum_{k=1}^{n} P_k A_k \, |T_k|^{2\alpha} \, A_k^* \right) \right]^{1/2} \tag{4.56}$$

$$\times \left[\operatorname{tr} \left(\sum_{k=1}^{n} P_k A_k \, |T_k^*|^{2(1-\alpha)} \, A_k^* \right) \right]^{1/2} - \left| \operatorname{tr} \left(\sum_{k=1}^{n} P_k A_k T_k A_k^* \right) \right|,$$

where $\mathbf{T} = (T_1, ..., T_n) \in \mathcal{B}^{(n)}(H)$, $\mathbf{A} = (A_1, ..., A_n) \in \mathcal{B}_2^{(n)}(H)$ and $\alpha \in [0, 1]$.

Utilising a similar argument to the one in Theorem 4.12 we can state:

Proposition 4.16 (Dragomir [8]) *Let* $\mathbf{T} = (T_1, ..., T_n) \in \mathcal{B}^{(n)}(H)$, $\mathbf{A} = (A_1, ..., A_n)$ $\in \mathcal{B}_2^{(n)}(H)$ *and* $\alpha \in [0, 1]$.

(i) For any $\mathbf{P}, \mathbf{Q} \in \mathcal{B}_+^{(n)}(H)$ *we have*

$$\sigma_{\mathbf{A},\mathbf{T},\alpha}(\mathbf{P} + \mathbf{Q}) \geq \sigma_{\mathbf{A},\mathbf{T},\alpha}(\mathbf{P}) + \sigma_{\mathbf{A},\mathbf{T},\alpha}(\mathbf{Q}) \, (\geq 0), \tag{4.57}$$

namely, $\sigma_{\mathbf{A},\mathbf{T},\alpha}$ *is a superadditive functional on* $\mathcal{B}_+^{(n)}(H)$;

(ii) For any $\mathbf{P}, \mathbf{Q} \in \mathcal{B}_+^{(n)}(H)$ *with* $\mathbf{P} \geq \mathbf{Q}$, *namely* $P_k \geq Q_k$ *for all* $k \in \{1, ..., n\}$ *we have*

$$\sigma_{\mathbf{A},\mathbf{T},\alpha}(\mathbf{P}) \geq \sigma_{\mathbf{A},\mathbf{T},\alpha}(\mathbf{Q}) \, (\geq 0), \tag{4.58}$$

namely, $\sigma_{\mathbf{A},\mathbf{B}}$ *is a monotonic nondecreasing functional on* $\mathcal{B}_+^{(n)}(H)$;

(iii) If $\mathbf{P}, \mathbf{Q} \in \mathcal{B}_+^{(n)}(H)$ *and there exist the constants* $M > m > 0$ *such that* $M\mathbf{Q} \geq \mathbf{P} \geq m\mathbf{Q}$ *then*

$$M\sigma_{\mathbf{A},\mathbf{T},\alpha}(\mathbf{Q}) \geq \sigma_{\mathbf{A},\mathbf{T},\alpha}(\mathbf{P}) \geq m\sigma_{\mathbf{A},\mathbf{T},\alpha}(\mathbf{Q}) \, (\geq 0). \tag{4.59}$$

If $\mathbf{P} = (p_1 1_H, ..., p_n 1_H)$ with $p_k \geq 0$, $k \in \{1, ..., n\}$ then the functional of real nonnegative weights $\mathbf{p} = (p_1, ..., p_n)$ defined by

$$\sigma_{A,T,\alpha}(\mathbf{p}) := \left[\operatorname{tr}\left(\sum_{k=1}^{n} p_k |A_k|^2 |T_k|^{2\alpha}\right)\right]^{1/2} \tag{4.60}$$

$$\times \left[\operatorname{tr}\left(\sum_{k=1}^{n} p_k |A_k|^2 |T_k^*|^{2(1-\alpha)}\right)\right]^{1/2} - \left|\operatorname{tr}\left(\sum_{k=1}^{n} p_k |A_k|^2 T_k\right)\right|$$

has the same properties as in Theorem 4.12.

Moreover, we have the simple bounds

$$\max_{k \in \{1,\dots,n\}} \{p_k\} \left(\left[\operatorname{tr}\left(\sum_{k=1}^{n} |A_k|^2 |T_k|^{2\alpha}\right)\right]^{1/2} \tag{4.61}\right.$$

$$\times \left[\operatorname{tr}\left(\sum_{k=1}^{n} |A_k|^2 |T_k^*|^{2(1-\alpha)}\right)\right]^{1/2} - \left|\operatorname{tr}\left(\sum_{k=1}^{n} p_k |A_k|^2 T_k\right)\right|\right)$$

$$\geq \left[\operatorname{tr}\left(\sum_{k=1}^{n} p_k |A_k|^2 |T_k|^{2\alpha}\right)\right]^{1/2} \left[\operatorname{tr}\left(\sum_{k=1}^{n} p_k |A_k|^2 |T_k^*|^{2(1-\alpha)}\right)\right]^{1/2}$$

$$- \left|\operatorname{tr}\left(\sum_{k=1}^{n} p_k |A_k|^2 T_k\right)\right|$$

$$\geq \min_{k \in \{1,\dots,n\}} \{p_k\} \left(\left[\operatorname{tr}\left(\sum_{k=1}^{n} |A_k|^2 |T_k|^{2\alpha}\right)\right]^{1/2}\right.$$

$$\times \left[\operatorname{tr}\left(\sum_{k=1}^{n} |A_k|^2 |T_k^*|^{2(1-\alpha)}\right)\right]^{1/2} - \left|\operatorname{tr}\left(\sum_{k=1}^{n} p_k |A_k|^2 T_k\right)\right|\right).$$

4.5 Inequalities for Power Series of Operators

We have the following trace inequalities:

Theorem 4.17 (Dragomir [8]) *Let* $f(\lambda) := \sum_{n=1}^{\infty} \alpha_n \lambda^n$ *be a power series with complex coefficients and convergent on the open disk* $D(0, R)$, $R > 0$. *Let* $N \in \mathcal{B}(H)$ *be a normal operator. If for some* $\alpha \in (0, 1)$ *we have* $|N|^{2\alpha}$, $|N|^{2(1-\alpha)} \in \mathcal{B}_1(H)$ *with* $\operatorname{tr}\left(|N|^{2\alpha}\right)$, $\operatorname{tr}\left(|N|^{2(1-\alpha)}\right) < R$, *then we have the inequality*

$$|\operatorname{tr}(f(N))|^2 \leq \operatorname{tr}\left(f_a\left(|N|^{2\alpha}\right)\right) \operatorname{tr}\left(f_a\left(|N|^{2(1-\alpha)}\right)\right). \tag{4.62}$$

Proof Since N is a normal operator, then for any natural number $k \geq 1$ we have $\left|N^k\right|^{2\alpha} = |N|^{2\alpha k}$ and $\left|N^k\right|^{2(1-\alpha)} = |N|^{2(1-\alpha)k}$.

By the generalized triangle inequality for the modulus we have for $n \geq 2$

$$\left| \operatorname{tr} \left(\sum_{k=1}^{n} \alpha_k N^k \right) \right| = \left| \sum_{k=1}^{n} \alpha_k \operatorname{tr} \left(N^k \right) \right| \leq \sum_{k=1}^{n} |\alpha_k| \left| \operatorname{tr} \left(N^k \right) \right|. \qquad (4.63)$$

If for some $\alpha \in (0, 1)$ we have $|N|^{2\alpha}$, $|N|^{2(1-\alpha)} \in \mathcal{B}_1(H)$, then by Corollary 4.7 we have $N \in \mathcal{B}_1(H)$. Now, since N, $|N|^{2\alpha}$, $|N|^{2(1-\alpha)} \in \mathcal{B}_1(H)$ then any natural power of these operators belong to $\mathcal{B}_1(H)$ and by (4.24) we have

$$\left| \operatorname{tr} \left(N^k \right) \right|^2 \leq \operatorname{tr} \left(|N|^{2\alpha k} \right) \operatorname{tr} \left(|N|^{2(1-\alpha)k} \right), \qquad (4.64)$$

for any natural number $k \geq 1$.

Making use of (4.64) we have

$$\sum_{k=1}^{n} |\alpha_k| \left| \operatorname{tr} \left(N^k \right) \right| \leq \sum_{k=1}^{n} |\alpha_k| \left(\operatorname{tr} \left(|N|^{2\alpha k} \right) \right)^{1/2} \left(\operatorname{tr} \left(|N|^{2(1-\alpha)k} \right) \right)^{1/2}. \qquad (4.65)$$

Utilising the weighted Cauchy–Bunyakovsky–Schwarz inequality for sums we also have

$$\sum_{k=1}^{n} |\alpha_k| \left(\operatorname{tr} \left(|N|^{2\alpha k} \right) \right)^{1/2} \left(\operatorname{tr} \left(|N|^{2(1-\alpha)k} \right) \right)^{1/2} \qquad (4.66)$$

$$\leq \left[\sum_{k=1}^{n} |\alpha_k| \left(\left(\operatorname{tr} \left(|N|^{2\alpha k} \right) \right)^{1/2} \right)^2 \right]^{1/2}$$

$$\times \left[\sum_{k=1}^{n} |\alpha_k| \left(\left(\operatorname{tr} \left(|N|^{2(1-\alpha)k} \right) \right)^{1/2} \right)^2 \right]^{1/2}$$

$$= \left[\sum_{k=1}^{n} |\alpha_k| \operatorname{tr} \left(|N|^{2\alpha k} \right) \right]^{1/2} \left[\sum_{k=1}^{n} |\alpha_k| \operatorname{tr} \left(|N|^{2(1-\alpha)k} \right) \right]^{1/2}.$$

Making use of (4.63), (4.65) and (4.66) we get the inequality

$$\left| \operatorname{tr} \left(\sum_{k=1}^{n} \alpha_k N^k \right) \right|^2 \leq \operatorname{tr} \left(\sum_{k=1}^{n} |\alpha_k| |N|^{2\alpha k} \right) \operatorname{tr} \left(\sum_{k=1}^{n} |\alpha_k| |N|^{2(1-\alpha)k} \right) \qquad (4.67)$$

for any $n \geq 2$.

Due to the fact that $\operatorname{tr} \left(|N|^{2\alpha} \right)$, $\operatorname{tr} \left(|N|^{2(1-\alpha)} \right) < R$ it follows by (4.24) that $\operatorname{tr} (|N|) < R$ and the operator series

$$\sum_{k=1}^{\infty} \alpha_k N^k, \quad \sum_{k=1}^{\infty} |\alpha_k| \, |N|^{2\alpha k} \quad \text{and} \quad \sum_{k=1}^{\infty} |\alpha_k| \, |N|^{2(1-\alpha)k}$$

are convergent in the Banach space $\mathcal{B}_1(H)$.

Taking the limit over $n \to \infty$ in (4.67) and using the continuity of the tr$(\cdot)$ on $\mathcal{B}_1(H)$ we deduce the desired result (4.62). $\blacksquare$

Example 4.18 (a) If we take in $f(\lambda) = (1 \pm \lambda)^{-1} - 1 = \mp \lambda \left((1 \pm \lambda)^{-1}\right)$, $|\lambda| < 1$ then we get from (4.62) the inequality

$$\left| \text{tr} \left(N \left((1_H \pm N)^{-1}\right)\right) \right|^2 \tag{4.68}$$
$$\leq \text{tr} \left(|N|^{2\alpha} \left(1_H - |N|^{2\alpha}\right)^{-1}\right) \text{tr} \left(|N|^{2(1-\alpha)} \left(1_H - |N|^{2(1-\alpha)}\right)^{-1}\right),$$

provided that $N \in \mathcal{B}(H)$ is a normal operator and for $\alpha \in (0,1)$ we have $|N|^{2\alpha}$, $|N|^{2(1-\alpha)} \in \mathcal{B}_1(H)$ with tr$\left(|N|^{2\alpha}\right)$, tr$\left(|N|^{2(1-\alpha)}\right) < 1$.

(b) If we take in (4.62) $f(\lambda) = \exp(\lambda) - 1$, $\lambda \in \mathbb{C}$ then we get the inequality

$$|\text{tr}(\exp(N) - 1_H)|^2 \leq \text{tr}\left(\exp\left(|N|^{2\alpha}\right) - 1_H\right) \text{tr}\left(\exp\left(|N|^{2(1-\alpha)}\right) - 1_H\right), \quad (4.69)$$

provided that $N \in \mathcal{B}(H)$ is a normal operator and for $\alpha \in (0,1)$ we have $|N|^{2\alpha}$, $|N|^{2(1-\alpha)} \in \mathcal{B}_1(H)$.

The following result also holds:

Theorem 4.19 (Dragomir [8]) *Let $f(\lambda) := \sum_{n=0}^{\infty} \alpha_n \lambda^n$ be a power series with complex coefficients and convergent on the open disk $D(0, R)$, $R > 0$. If $T \in \mathcal{B}(H)$, $A \in \mathcal{B}_2(H)$ are normal operators that double commute, i.e. $TA = AT$ and $TA^* = A^*T$ and tr$\left(|A|^2 \, |T|^{2\alpha}\right)$, tr$\left(|A|^2 \, |T|^{2(1-\alpha)}\right) < R$ for some $\alpha \in [0,1]$, then*

$$\left| \text{tr} \left(f \left(|A|^2 T\right)\right) \right|^2 \leq \text{tr} \left(f_a \left(|A|^2 \, |T|^{2\alpha}\right)\right) \text{tr} \left(f_a \left(|A|^2 \, |T|^{2(1-\alpha)}\right)\right). \tag{4.70}$$

Proof From the inequality (4.55) we have

$$\left| \text{tr} \left(\sum_{k=0}^{n} \alpha_k \, |A^k|^2 \, T^k \right) \right|^2 \tag{4.71}$$
$$\leq \text{tr} \left(\sum_{k=0}^{n} |\alpha_k| \, |A^k|^2 \, |T^k|^{2\alpha}\right) \text{tr} \left(\sum_{k=0}^{n} |\alpha_k| \, |A^k|^2 \, |T^k|^{2(1-\alpha)}\right).$$

Since A and T are normal operators, then $|A^k|^2 = |A|^{2k}$, $|T^k|^{2\alpha} = |T|^{2\alpha k}$ and $|T^k|^{2(1-\alpha)} = |T|^{2(1-\alpha)k}$ for any natural number $k \geq 0$ and $\alpha \in [0,1]$.

Since T and A double commute, then is easy to see that

$$|A|^{2k} T^k = \left(|A|^2 T\right)^k, \quad |A|^{2k} |T|^{2\alpha k} = \left(|A|^2 |T|^{2\alpha}\right)^k$$

and

$$|A|^{2k} |T|^{2(1-\alpha)k} = \left(|A|^2 |T|^{2(1-\alpha)}\right)^k$$

for any natural number $k \geq 0$ and $\alpha \in [0, 1]$.

Therefore (4.71) is equivalent to

$$\left| \operatorname{tr} \left(\sum_{k=0}^{n} \alpha_k \left(|A|^2 T\right)^k \right) \right|^2 \tag{4.72}$$

$$\leq \operatorname{tr} \left(\sum_{k=0}^{n} |\alpha_k| \left(|A|^2 |T|^{2\alpha}\right)^k \right) \operatorname{tr} \left(\sum_{k=0}^{n} |\alpha_k| \left(|A|^2 |T|^{2(1-\alpha)}\right)^k \right),$$

for any natural number $n \geq 1$ and $\alpha \in [0, 1]$.

Due to the fact that $\operatorname{tr} \left(|A|^2 |T|^{2\alpha}\right)$, $\operatorname{tr} \left(|A|^2 |T|^{2(1-\alpha)}\right) < R$ it follows by (4.55) for $n = 1$ that $\operatorname{tr} \left(|A|^2 T\right) < R$ and the operator series

$$\sum_{k=1}^{\infty} \alpha_k N^k, \quad \sum_{k=1}^{\infty} |\alpha_k| |N|^{2\alpha k} \text{ and } \sum_{k=1}^{\infty} |\alpha_k| |N|^{2(1-\alpha)k}$$

are convergent in the Banach space $\mathcal{B}_1(H)$.

Taking the limit over $n \to \infty$ in (4.72) and using the continuity of the $\operatorname{tr}(\cdot)$ on $\mathcal{B}_1(H)$ we deduce the desired result (4.70). $\blacksquare$

Example 4.20 (a) If we take $f(\lambda) = (1 \pm \lambda)^{-1}$, $|\lambda| < 1$ then we get from (4.70) the inequality

$$\left| \operatorname{tr} \left(\left(1_H \pm |A|^2 T\right)^{-1} \right) \right|^2 \tag{4.73}$$

$$\leq \operatorname{tr} \left(\left(1_H - |A|^2 |T|^{2\alpha}\right)^{-1} \right) \operatorname{tr} \left(\left(1_H - |A|^2 |T|^{2(1-\alpha)}\right)^{-1} \right),$$

provided that $T \in \mathcal{B}(H)$, $A \in \mathcal{B}_2(H)$ are normal operators that double commute and $\operatorname{tr} \left(|A|^2 |T|^{2\alpha}\right)$, $\operatorname{tr} \left(|A|^2 |T|^{2(1-\alpha)}\right) < 1$ for $\alpha \in [0, 1]$.

(b) If we take in (4.70) $f(\lambda) = \exp(\lambda)$, $\lambda \in \mathbb{C}$ then we get the inequality

$$\left| \operatorname{tr} \left(\exp \left(|A|^2 T\right) \right) \right|^2 \leq \operatorname{tr} \left(\exp \left(|A|^2 |T|^{2\alpha}\right) \right) \operatorname{tr} \left(\exp \left(|A|^2 |T|^{2(1-\alpha)}\right) \right), \tag{4.74}$$

provided that $T \in \mathcal{B}(H)$ and $A \in \mathcal{B}_2(H)$ are normal operators that double commute and $\alpha \in [0, 1]$.

Theorem 4.21 (Dragomir [8]) *Let* $f(z) := \sum_{j=0}^{\infty} p_j z^j$ *and* $g(z) := \sum_{j=0}^{\infty} q_j z^j$ *be two power series with nonnegative coefficients and convergent on the open disk* $D(0, R)$, $R > 0$. *If* $T \in \mathcal{B}(H)$, $A \in \mathcal{B}_2(H)$ *are normal operators that double commute and* $\mathrm{tr}\left(|A|^2 |T|^{2\alpha}\right)$, $\mathrm{tr}\left(|A|^2 |T|^{2(1-\alpha)}\right) < R$ *for* $\alpha \in [0, 1]$, *then*

$$\left[\mathrm{tr}\left(f\left(|A|^2 |T|^{2\alpha}\right) + g\left(|A|^2 |T|^{2\alpha}\right)\right)\right]^{1/2} \tag{4.75}$$
$$\times \left[\mathrm{tr}\left(f\left(|A|^2 |T|^{2(1-\alpha)}\right) + g\left(|A|^2 |T|^{2(1-\alpha)}\right)\right)\right]^{1/2}$$
$$- \left|\mathrm{tr}\left(f\left(|A|^2 T\right) + g\left(|A|^2 T\right)\right)\right|$$
$$\geq \left[\mathrm{tr}\left(f\left(|A|^2 |T|^{2\alpha}\right)\right)\right]^{1/2} \left[\mathrm{tr}\left(f\left(|A|^2 |T|^{2(1-\alpha)}\right)\right)\right]^{1/2}$$
$$- \left|\mathrm{tr}\left(f\left(|A|^2 T\right)\right)\right|$$
$$+ \left[\mathrm{tr}\left(g\left(|A|^2 |T|^{2\alpha}\right)\right)\right]^{1/2} \left[\mathrm{tr}\left(g\left(|A|^2 |T|^{2(1-\alpha)}\right)\right)\right]^{1/2}$$
$$- \left|\mathrm{tr}\left(g\left(|A|^2 T\right)\right)\right| (\geq 0).$$

Moreover, if $p_j \geq q_j$ *for any* $j \in \mathbb{N}$, *then, with the above assumptions on* T *and* A, *we have*

$$\left[\mathrm{tr}\left(f\left(|A|^2 |T|^{2\alpha}\right)\right)\right]^{1/2} \left[\mathrm{tr}\left(f\left(|A|^2 |T|^{2(1-\alpha)}\right)\right)\right]^{1/2} \tag{4.76}$$
$$- \left|\mathrm{tr}\left(f\left(|A|^2 T\right)\right)\right|$$
$$\geq \left[\mathrm{tr}\left(g\left(|A|^2 |T|^{2\alpha}\right)\right)\right]^{1/2} \left[\mathrm{tr}\left(g\left(|A|^2 |T|^{2(1-\alpha)}\right)\right)\right]^{1/2}$$
$$- \left|\mathrm{tr}\left(g\left(|A|^2 T\right)\right)\right| (\geq 0).$$

The proof follows in a similar way to the proof of Theorem 4.19 by making use of the superadditivity and monotonicity properties of the functional $\sigma_{A,T,\alpha}(\cdot)$. We omit the details.

Example 4.22 Now, observe that if we take

$$f(\lambda) = \sinh \lambda = \sum_{n=0}^{\infty} \frac{1}{(2n+1)!} \lambda^{2n+1}$$

and

$$g(\lambda) = \cosh \lambda = \sum_{n=0}^{\infty} \frac{1}{(2n)!} \lambda^{2n}$$

then

$$f(\lambda) + g(\lambda) = \exp \lambda = \sum_{n=0}^{\infty} \frac{1}{n!} \lambda^n$$

for any $\lambda \in \mathbb{C}$.

If $T \in \mathcal{B}(H)$, $A \in \mathcal{B}_2(H)$ are normal operators that double commute and $\alpha \in [0, 1]$, then by (4.75) we have

$$\left[\operatorname{tr}\left(\exp\left(|A|^2\,|T|^{2\alpha}\right)\right)\right]^{1/2}\left[\operatorname{tr}\left(\exp\left(|A|^2\,|T|^{2(1-\alpha)}\right)\right)\right]^{1/2} \qquad (4.77)$$
$$-\left|\operatorname{tr}\left(\exp\left(|A|^2\,T\right)\right)\right|$$
$$\geq \left[\operatorname{tr}\left(\sinh\left(|A|^2\,|T|^{2\alpha}\right)\right)\right]^{1/2}\left[\operatorname{tr}\left(\sinh\left(|A|^2\,|T|^{2(1-\alpha)}\right)\right)\right]^{1/2}$$
$$-\left|\operatorname{tr}\left(\sinh\left(|A|^2\,T\right)\right)\right|$$
$$+\left[\operatorname{tr}\left(\cosh\left(|A|^2\,|T|^{2\alpha}\right)\right)\right]^{1/2}\left[\operatorname{tr}\left(\cosh\left(|A|^2\,|T|^{2(1-\alpha)}\right)\right)\right]^{1/2}$$
$$-\left|\operatorname{tr}\left(\cosh\left(|A|^2\,T\right)\right)\right| \; (\geq 0).$$

Now, consider the series $\frac{1}{1-\lambda} = \sum_{n=0}^{\infty} \lambda^n$, $\lambda \in D(0,1)$ and $\ln \frac{1}{1-\lambda} = \sum_{n=1}^{\infty} \frac{1}{n}\lambda^n$, $\lambda \in D(0,1)$ and define $p_n = 1, n \geq 0$, $q_0 = 0$, $q_n = \frac{1}{n}, n \geq 1$, then we observe that for any $n \geq 0$ we have $p_n \geq q_n$.

If $T \in \mathcal{B}(H)$, $A \in \mathcal{B}_2(H)$ are normal operators that double commute, $\alpha \in [0, 1]$ and $\operatorname{tr}\left(|A|^2\,|T|^{2\alpha}\right)$, $\operatorname{tr}\left(|A|^2\,|T|^{2(1-\alpha)}\right) < 1$, then by (4.76) we have

$$\left[\operatorname{tr}\left(\left(1_H - |A|^2\,|T|^{2\alpha}\right)^{-1}\right)\right]^{1/2}\left[\operatorname{tr}\left(\left(1_H - |A|^2\,|T|^{2(1-\alpha)}\right)^{-1}\right)\right]^{1/2} \qquad (4.78)$$
$$-\left|\operatorname{tr}\left(\left(1_H - |A|^2\,T\right)^{-1}\right)\right|$$
$$\geq \left[\operatorname{tr}\left(\ln\left(1_H - |A|^2\,|T|^{2\alpha}\right)^{-1}\right)\right]^{1/2}\left[\operatorname{tr}\left(\ln\left(1_H - |A|^2\,|T|^{2(1-\alpha)}\right)^{-1}\right)\right]^{1/2}$$
$$-\left|\operatorname{tr}\left(\ln\left(1_H - |A|^2\,T\right)^{-1}\right)\right| \; (\geq 0).$$

Chapter 5
Integral Inequalities

In this chapter, after recalling some fundamental facts on Bochner integral for measurable functions with values in Banach spaces, we provide an integral version of Kato's inequality. Several *Norm* and *Numerical Radius* inequalities with applications for the *Operator Exponential* are also given.

5.1 Some Facts on Bochner Integral

Let $\mathcal{F}(B; E, \mathcal{A}, \mu)$ be the linear space of functions $x(t)$, $t \in E$, with values in a real or complex Banach space B, given on a measurable space $(E, \mathcal{A}, \mu)$ endowed with a countably-additive scalar measure μ on a σ-algebra $\mathcal{A}$ of subsets of E.

A function $x_0 \in \mathcal{F}$ is called *simple* if can be defined as, see [32]

$$
x_0(t) := \begin{cases} x_i \in B, \, t \in A_i \in \mathcal{A}, \, \mu(A_i) < \infty, \, i \in \{1, ..., n\} \\ \qquad A_k \cap A_j = \emptyset, \, k \neq j, \, k, \, j \in \{1, ..., n\}, \\ 0, \qquad t \in E \setminus \cup_{i=1}^{n} A_i, \, n \in \mathbb{N}. \end{cases}
$$

A function $x \in \mathcal{F}$ is called *strongly measurable* if there exists a sequence $\{x_n\}$ of simple functions with $\|x_n - x\| \to 0$ almost-everywhere with respect to the measure μ on E. As a consequence of this, the scalar function $\|x\|$ is $\mathcal{A}$-measurable.

For the simple function $x_0 \in \mathcal{F}$ as above we define the integral by

$$
\int_E x_0(t) \, d\mu(t) := \sum_{i=1}^{n} x_i \mu(A_i).
$$

A function $x \in \mathcal{F}$ is said to be *Bochner integrable* if it is strongly measurable and if for some approximating sequence $\{x_n\}$ of simple functions we have

$$
\lim_{n \to \infty} \int_E \|x(t) - x_n(t)\| \, d\mu(t) = 0.
$$

© The Author(s), under exclusive license to Springer Nature Switzerland AG 2019
S. S. Dragomir, *Kato's Type Inequalities for Bounded Linear Operators in Hilbert Spaces*, SpringerBriefs in Mathematics,
https://doi.org/10.1007/978-3-030-17459-0_5

The *Bochner integral* of such a function over a set $A \in \mathcal{A}$ is defined as

$$\int_A x(t) \, d\mu(t) = \lim_{n \to \infty} \int_E \chi_A(t) \, x_n(t) \, d\mu(t),$$

where χ_A is the *characteristic function* of A, and the limit is understood in the sense of strong convergence in the Banach space E. This limit exists, and is independent of the choice of the approximation sequence of simple functions.

It is well-known that, for a strongly-measurable function to be Bochner integrable it is necessary and sufficient for the norm of this function to be integrable, i.e.

$$\int_A \|x(t)\| \, d\mu(t) < \infty.$$

The set of Bochner-integrable functions forms a vector subspace $\mathcal{L}(B; E, \mathcal{A}, \mu)$ of $\mathcal{F}(B; E, \mathcal{A}, \mu)$, and the Bochner integral is a linear operator on this subspace.

Some fundamental properties of Bochner integrals are as follows [32],

1. For any $x \in \mathcal{L}(B; E, \mathcal{A}, \mu)$ we have the norm inequality

$$\left\| \int_A x(t) \, d\mu(t) \right\| \leq \int_A \|x(t)\| \, d\mu(t).$$

2. Bochner integral is a countably-additive μ-absolutely continuous set-function on the σ-algebra $\mathcal{A}$, i.e.

$$\int_{\cup_{i=1}^{\infty} A_i} x(t) \, d\mu(t) = \sum_{i=1}^{\infty} \int_{A_i} x(t) \, d\mu(t)$$

if $A_i \in \mathcal{A}, \mu(A_i) < \infty, i \in \mathbb{N}, A_k \cap A_j = \emptyset, k \neq j, k, j \in \mathbb{N}$, and

$$\left\| \int_A x(t) \, d\mu(t) \right\| \to 0 \text{ if } \mu(A) \to 0,$$

uniformly over $A \in \mathcal{A}$.

3. If $x_n \in F, x_n \to x$ almost-everywhere with respect to the measure μ on $A \in \mathcal{A}$, if $\|x_n\| \leq f$ almost-everywhere with respect to μ on A, and if $\int_A f(t) \, d\mu(t) < \infty$, then $x \in \mathcal{L}(B; E, \mathcal{A}, \mu)$ and

$$\int_A x_n(t) \, d\mu(t) \to \int_A x(t) \, d\mu(t).$$

4. The space is complete with respect to the norm

$$\|x\| := \int_A \|x(t)\| \, d\mu(t).$$

5. If T is a closed linear operator from a Banach space X into a Banach space Y and if $x \in \mathcal{L}(X; E, \mathcal{A}, \mu)$ and $Tx \in \mathcal{L}(Y; E, \mathcal{A}, \mu)$, then

$$\int_A Tx(t) \, d\mu(t) = T\left(\int_A x(t) \, d\mu(t)\right).$$

If T is bounded, the condition $Tx \in \mathcal{L}(Y; E, \mathcal{A}, \mu)$ is automatically satisfied.

5.2 Applications of Kato's Inequality

In this section we consider a measurable space $(E, \mathcal{A}, \mu)$ and operator-valued μ-measurable functions $E \ni t \longmapsto V_t \in \mathcal{B}(H)$, where $\mathcal{B}(H)$ denotes the Banach algebra of all bounded linear operators on a complex Hilbert space $(H; \langle \cdot, \cdot \rangle)$. We can define the adjoint function by $V_t^* := (V_t)^*$ and the modulus function by $|V|_t := \sqrt{(V_t)^* V_t}, t \in E$.

Theorem 5.1 (Dragomir [5]) *Let $V_{(\cdot)} : E \to \mathcal{B}(H)$ and $p : E \to [0, \infty)$ be μ-measurable functions on E and such that $p\,|V|_{(\cdot)}^2$ and $p\,|V^*|_{(\cdot)}^2$ are Bochner integrable on E. Then we have the inequality*

$$\int_E p(t) \, |\langle V_t x, y \rangle|^2 \, d\mu(t) \tag{5.1}$$

$$\leq \left\langle \left(\int_E p(t) \, |V|_t^2 \, d\mu(t)\right) x, x \right\rangle^\alpha \left\langle \left(\int_E p(t) \, |V^*|_t^2 \, d\mu(t)\right) y, y \right\rangle^{1-\alpha}$$

for any $x, y \in H$ with $\|x\| = \|y\| = 1$ and $\alpha \in [0, 1]$.

Proof Let $t \in E$. If we write Kato's inequality for the operator V_t we have

$$|\langle V_t x, y \rangle|^2 \leq \langle |V|_t^{2\alpha} x, x \rangle \langle |V^*|_t^{2(1-\alpha)} y, y \rangle \tag{5.2}$$

for any $x, y \in H$.

By Hölder-McCarthy inequality $\langle P^r x, x \rangle \leq \langle Px, x \rangle^r$ that holds for any positive operator P, for any $x \in H$ with $\|x\| = 1$ and any power $r \in (0, 1)$ we have

$$\langle |V|_t^{2\alpha} x, x \rangle \leq \langle |V_t|^2 x, x \rangle^\alpha \tag{5.3}$$

and

$$\langle |V^*|_t^{2(1-\alpha)} y, y \rangle \leq \langle |V^*|_t^2 y, y \rangle^{1-\alpha} \tag{5.4}$$

for any $x, y \in H$ with $\|x\| = \|y\| = 1$.

On making use of (5.2)–(5.4) we get

$$|\langle V_t x, y \rangle|^2 \leq \left\langle |V|^2_t x, x \right\rangle^\alpha \left\langle |V^*|^2_t y, y \right\rangle^{1-\alpha} \tag{5.5}$$

for any $x, y \in H$ with $\|x\| = \|y\| = 1$ and $t \in E$.

Now, if we multiply the inequality (5.5) by $p(t) \geq 0$, integrate over $d\mu(\cdot)$ on E and use the scalar weighted Hölder inequality we have

$$\int_E p(t) |\langle V_t x, y \rangle|^2 \, d\mu(t)$$

$$\leq \int_E p(t) \left\langle |V|^2_t x, x \right\rangle^\alpha \left\langle |V^*|^2_t y, y \right\rangle^{1-\alpha} \, d\mu(t)$$

$$\leq \left(\int_E p(t) \left[\langle |V|^2_t x, x \rangle^\alpha \right]^{1/\alpha} \, d\mu(t) \right)^\alpha$$

$$\times \left(\int_E p(t) \left[\langle |V^*|^2_t y, y \rangle^{1-\alpha} \right]^{1/(1-\alpha)} \, d\mu(t) \right)^{1-\alpha}$$

$$= \left(\int_E p(t) \left\langle |V|^2_t x, x \right\rangle d\mu(t) \right)^\alpha \left(\int_E p(t) \left\langle |V^*|^2_t y, y \right\rangle d\mu(t) \right)^{1-\alpha}$$

$$= \left\langle \left(\int_E p(t) |V|^2_t \, d\mu(t) \right) x, x \right\rangle^\alpha \left\langle \left(\int_E p(t) |V^*|^2_t \, d\mu(t) \right) y, y \right\rangle^{1-\alpha}$$

for any $x, y \in H$ with $\|x\| = \|y\| = 1$, and the proof is complete. ∎

Remark 5.2 The inequality (5.1) becomes for $y = x$ the following simpler result that is useful for deriving numerical radius inequalities:

$$\int_E p(t) |\langle V_t x, x \rangle|^2 \, d\mu(t) \tag{5.6}$$

$$\leq \left\langle \left(\int_E p(t) |V|^2_t \, d\mu(t) \right) x, x \right\rangle^\alpha \left\langle \left(\int_E p(t) |V^*|^2_t \, d\mu(t) \right) x, x \right\rangle^{1-\alpha}$$

$$\leq \left\langle \left(\int_E p(t) \left[\alpha |V|^2_t + (1-\alpha) |V^*|^2_t \right] d\mu(t) \right) x, x \right\rangle$$

for any $x \in H$ with $\|x\| = 1$.

Remark 5.3 In addition to the assumptions of Theorem 5.1, if the values of the function $V_{(\cdot)}$ are normal operators for μ-almost every (a.e.) $t \in E$, i.e., $|V|^2_t = |V^*|^2_t$ for μ-a.e. $t \in E$ we have

$$\int_E p(t) |\langle V_t x, y \rangle|^2 \, d\mu(t) \tag{5.7}$$

$$\leq \left\langle \left(\int_E p\,(t)\,|V|_t^2\,d\mu\,(t) \right) x, x \right\rangle^\alpha \left\langle \left(\int_E p\,(t)\,|V|_t^2\,d\mu\,(t) \right) y, y \right\rangle^{1-\alpha}$$

for any $x, y \in H$ with $\|x\| = \|y\| = 1$ and $\alpha \in [0, 1]$.

This inequality implies the simpler result

$$\int_E p\,(t)\,|\langle V_t x, x \rangle|^2\,d\mu\,(t) \leq \left\langle \left(\int_E p\,(t)\,|V|_t^2\,d\mu\,(t) \right) x, x \right\rangle \tag{5.8}$$

for any $x \in H$ with $\|x\| = 1$.

From a different perspective, we can state the following result as well:

Theorem 5.4 (Dragomir [5]) *Let $V_{(\cdot)} : E \to \mathcal{B}(H)$ and $p : E \to [0, \infty)$ be μ-measurable functions on E and such that $p\,|V|_{(\cdot)}^{2\alpha}$ and $p\,|V^*|_{(\cdot)}^{2(1-\alpha)}$ are Bochner integrable on E for some $\alpha \in [0, 1]$. Then we have the inequality*

$$\int_E p\,(t)\,|\langle V_t x, y \rangle|\,d\mu\,(t) \tag{5.9}$$

$$\leq \left\langle \left(\int_E p\,(t)\,|V|_t^{2\alpha}\,d\mu\,(t) \right) x, x \right\rangle^{1/2} \left\langle \left(\int_E p\,(t)\,|V^*|_t^{2(1-\alpha)}\,d\mu\,(t) \right) y, y \right\rangle^{1/2}$$

for any $x, y \in H$.

In particular, we have

$$\int_E p\,(t)\,|\langle V_t x, x \rangle|\,d\mu\,(t) \tag{5.10}$$

$$\leq \left\langle \left(\int_E p\,(t)\,|V|_t^{2\alpha}\,d\mu\,(t) \right) x, x \right\rangle^{1/2} \left\langle \left(\int_E p\,(t)\,|V^*|_t^{2(1-\alpha)}\,d\mu\,(t) \right) x, x \right\rangle^{1/2}$$

$$\leq \frac{1}{2} \left\langle \left(\int_E p\,(t)\,\left[|V|_t^{2\alpha} + |V^*|_t^{2(1-\alpha)} \right] d\mu\,(t) \right) x, x \right\rangle$$

for any $x \in H$.

Proof Let $t \in E$. If we write Kato's inequality for the operator V_t we have

$$|\langle V_t x, y \rangle| \leq \left\langle |V|_t^{2\alpha} x, x \right\rangle^{1/2} \left\langle |V^*|_t^{2(1-\alpha)} y, y \right\rangle^{1/2} \tag{5.11}$$

for any $x, y \in H$.

Now, by multiplying the inequality (5.11) with $p\,(t) \geq 0$, integrate over $d\mu\,(\cdot)$ on E and use the weighted Cauchy–Bunyakovsky–Schwarz integral inequality we get

$$\int_E p\,(t)\,|\langle V_t x, y \rangle|\,d\mu\,(t)$$

$$\leq \int_E p(t) \left\langle |V|_t^{2\alpha} x, x\right\rangle^{1/2} \left\langle |V^*|_t^{2(1-\alpha)} y, y\right\rangle^{1/2} d\mu(t)$$

$$\leq \left(\int_E p(t) \left\langle |V|_t^{2\alpha} x, x\right\rangle d\mu(t)\right)^{1/2} \left(\int_E p(t) \left\langle |V^*|_t^{2(1-\alpha)} y, y\right\rangle d\mu(t)\right)^{1/2}$$

$$= \left\langle \left(\int_E p(t) |V|_t^{2\alpha} d\mu(t)\right) x, x\right\rangle^{1/2} \left\langle \left(\int_E p(t) |V^*|_t^{2(1-\alpha)} d\mu(t)\right) y, y\right\rangle^{1/2}$$

for any $x, y \in H$.

The second inequality in (5.10) follows by the arithmetic mean-geometric mean inequality.

The proof is complete.   ∎

Remark 5.5 The symmetric case for powers, namely the case $\alpha = \frac{1}{2}$ in (5.9) is of interest since will produce the simpler result

$$\int_E p(t) |\langle V_t x, y\rangle| d\mu(t) \tag{5.12}$$

$$\leq \left\langle \left(\int_E p(t) |V|_t d\mu(t)\right) x, x\right\rangle^{1/2} \left\langle \left(\int_E p(t) |V^*|_t d\mu(t)\right) y, y\right\rangle^{1/2}$$

for any $x, y \in H$ and provided that $p|V|_{(\cdot)}$ and $p|V^*|_{(\cdot)}$ are Bochner integrable on E.

If in this inequality we take $y = x$, then we get

$$\int_E p(t) |\langle V_t x, x\rangle| d\mu(t) \tag{5.13}$$

$$\leq \left\langle \left(\int_E p(t) |V|_t d\mu(t)\right) x, x\right\rangle^{1/2} \left\langle \left(\int_E p(t) |V^*|_t d\mu(t)\right) x, x\right\rangle^{1/2}$$

for any $x \in H$.

Moreover, if the values of the function $V_{(\cdot)}$ are normal operators for μ-a.e. $t \in E$, then the inequality (5.12) becomes

$$\int_E p(t) |\langle V_t x, y\rangle| d\mu(t) \tag{5.14}$$

$$\leq \left\langle \left(\int_E p(t) |V|_t d\mu(t)\right) x, x\right\rangle^{1/2} \left\langle \left(\int_E p(t) |V|_t d\mu(t)\right) y, y\right\rangle^{1/2}$$

for any $x, y \in H$, while the inequality (5.13) becomes

$$\int_E p(t) |\langle V_t x, x\rangle| d\mu(t) \leq \left\langle \left(\int_E p(t) |V|_t d\mu(t)\right) x, x\right\rangle \tag{5.15}$$

for any $x \in H$.

5.3 Norm and Numerical Radius Inequalities

Let $p : E \to [0, \infty)$ be a μ-measurable function on E and such that $\int_E p(t) \, d\mu(t) = 1$. For $V_{(\cdot)} : E \to \mathcal{B}(H)$ a μ-measurable function on E and such that $p \, |V|^2_{(\cdot)}$ is Bochner integrable on E, we define the s-2-p-$semi$-$norm$ by

$$\|V_{(\cdot)}\|_{s,p,2} := \sup_{\|x\|=\|y\|=1} \left(\int_E p(t) \, |\langle V_t x, y \rangle|^2 \, d\mu(t) \right)^{1/2}$$

and the s-2-p-$semi$-$numerical\ radius$ by

$$w_{s,p,2} (V_{(\cdot)}) := \sup_{\|x\|=1} \left(\int_E p(t) \, |\langle V_t x, x \rangle|^2 \, d\mu(t) \right)^{1/2}.$$

If we consider the Banach space $\mathcal{L}_{2,p} (E, \mathcal{B}(H), \mu)$ of all functions $V_{(\cdot)} : E \to \mathcal{B}(H)$ that are μ-measurable on E and such that

$$\|V_{(\cdot)}\|_{p,2} := \left(\int_E p(t) \, \|V_t\|^2 \, d\mu(t) \right)^{1/2} < \infty,$$

we observe that $\|\cdot\|_{p,2}$ and $w(\cdot)_{p,2}$ defined on $\mathcal{L}_{2,p} (E, \mathcal{B}(H), \mu)$ are nonnegative, absolute homogeneous and satisfy the triangle inequality on this space.

If we consider the norm on $\mathcal{L}_{2,p} (E, \mathcal{B}(H), \mu)$ induced by the numerical radius on $\mathcal{B}(H)$, i.e.

$$w_{p,2} (V_{(\cdot)}) := \left(\int_E p(t) \, w^2 (V_t) \, d\mu(t) \right)^{1/2}$$

then by taking into account the well known numerical radius-norm inequalities

$$\frac{1}{2} \|T\| \le w(T) \le \|T\|, \quad T \in \mathcal{B}(H) \tag{5.16}$$

we observe that the norms $\|\cdot\|_{p,2}$ and $w_{p,2}(\cdot)$ will preserve the inequalities (5.16).

Utilising the properties of the supremum, we also observe that

$$\|V_{(\cdot)}\|_{s,p,2} \le \|V_{(\cdot)}\|_{p,2} \quad \text{and} \quad w_{s,p,2} (V_{(\cdot)}) \le w_{p,2} (V_{(\cdot)}) \tag{5.17}$$

for any $V_{(\cdot)} \in \mathcal{L}_{2,p} (E, \mathcal{B}(H), \mu)$.

We have the following result.

Theorem 5.6 (Dragomir [5]) *For any $V_{(\cdot)} \in \mathcal{L}_{2,p} (E, \mathcal{B}(H), \mu)$ and $\alpha \in [0, 1]$ we have the inequalities*

$$\left\| \int_E p\,(t)\,V_t d\mu\,(t) \right\|^2 \le \left\| V_{(\cdot)} \right\|_{s,p,2}^2 \tag{5.18}$$

$$\le \left\| \int_E p\,(t)\,|V|_t^2\,d\mu\,(t) \right\|^\alpha \left\| \int_E p\,(t)\,|V^*|_t^2\,d\mu\,(t) \right\|^{1-\alpha}$$

and

$$w^2 \left(\int_E p\,(t)\,V_t d\mu\,(t) \right) \tag{5.19}$$

$$\le w_{s,p,2}^2 \left(V_{(\cdot)} \right)$$

$$\le \begin{cases} \left\| \int_E p\,(t)\,|V|_t^2\,d\mu\,(t) \right\|^\alpha \left\| \int_E p\,(t)\,|V^*|_t^2\,d\mu\,(t) \right\|^{1-\alpha}, \\[2mm] \left\| \int_E p\,(t)\,\left[\alpha\,|V|_t^2 + (1-\alpha)\,|V^*|_t^2 \right] d\mu\,(t) \right\|. \end{cases}$$

Proof By the Cauchy–Bunyakovsky–Schwarz integral inequality and the inequality (5.1) we have

$$\left| \left\langle \left(\int_E p\,(t)\,V_t d\mu\,(t) \right) x, y \right\rangle \right|^2 \le \int_E p\,(t)\,|\langle V_t x, y \rangle|^2\,d\mu\,(t) \tag{5.20}$$

$$\le \left\langle \left(\int_E p\,(t)\,|V|_t^2\,d\mu\,(t) \right) x, x \right\rangle^\alpha \left\langle \left(\int_E p\,(t)\,|V^*|_t^2\,d\mu\,(t) \right) y, y \right\rangle^{1-\alpha}$$

for any $x, y \in H$ with $\|x\| = \|y\| = 1$ and $\alpha \in [0, 1]$.

Taking the supremum over $\|x\| = \|y\| = 1$ we have

$$\left[\sup_{\|x\|=\|y\|=1} \left| \left\langle \left(\int_E p\,(t)\,V_t d\mu\,(t) \right) x, y \right\rangle \right| \right]^2 \tag{5.21}$$

$$\le \sup_{\|x\|=\|y\|=1} \int_E p\,(t)\,|\langle V_t x, y \rangle|^2\,d\mu\,(t)$$

$$\le \left[\sup_{\|x\|=1} \left\langle \left(\int_E p\,(t)\,|V|_t^2\,d\mu\,(t) \right) x, x \right\rangle \right]^\alpha$$

$$\times \left[\sup_{\|y\|=1} \left\langle \left(\int_E p\,(t)\,|V^*|_t^2\,d\mu\,(t) \right) y, y \right\rangle \right]^{1-\alpha}$$

and since

$$\sup_{\|x\|=\|y\|=1} \left| \left\langle \left(\int_E p\,(t)\,V_t d\mu\,(t) \right) x, y \right\rangle \right| = \left\| \int_E p\,(t)\,V_t d\mu\,(t) \right\|,$$

$$\sup_{\|x\|=\|y\|=1} \int_E p\,(t)\,|\langle V_t x, y\rangle|^2\,d\mu\,(t) = \big\|V_{(\cdot)}\big\|^2_{s,p,2},$$

and

$$\sup_{\|x\|=1} \left\langle \left(\int_E p\,(t)\,|V|^2_t\,d\mu\,(t)\right) x, x\right\rangle = w\left(\int_E p\,(t)\,|V|^2_t\,d\mu\,(t)\right)$$

$$= \left\|\int_E p\,(t)\,|V|^2_t\,d\mu\,(t)\right\|$$

while

$$\sup_{\|y\|=1} \left\langle \left(\int_E p\,(t)\,|V^*|^2_t\,d\mu\,(t)\right) y, y\right\rangle = w\left(\int_E p\,(t)\,|V^*|^2_t\,d\mu\,(t)\right)$$

$$= \left\|\int_E p\,(t)\,|V^*|^2_t\,d\mu\,(t)\right\|$$

since the operators $\int_E p\,(t)\,|V|^2_t\,d\mu\,(t)$ and $\int_E p\,(t)\,|V^*|^2_t\,d\mu\,(t)$ are selfadjoint, then we deduce from (5.21) the desired result (5.18).

From the inequality (5.6) we also have

$$\left|\left\langle \int_E p\,(t)\,V_t x, x\right\rangle\right|^2 d\mu\,(t)$$

$$\leq \int_E p\,(t)\,|\langle V_t x, x\rangle|^2\,d\mu\,(t)$$

$$\leq \begin{cases} \left\langle \left(\int_E p\,(t)\,|V|^2_t\,d\mu\,(t)\right) x, x\right\rangle^\alpha \left\langle \left(\int_E p\,(t)\,|V^*|^2_t\,d\mu\,(t)\right) x, x\right\rangle^{1-\alpha}, \\ \left\langle \left(\int_E p\,(t)\,[\alpha\,|V|^2_t + (1-\alpha)\,|V^*|^2_t]\,d\mu\,(t)\right) x, x\right\rangle \end{cases}$$

for any $x \in H$ with $\|x\| = 1$.

Taking the supremum over $\|x\| = 1$ we deduce the desired inequality (5.19). The details are omitted. ∎

Remark 5.7 Since by the integral triangle inequality for the norm we have

$$\left\|\int_E p\,(t)\,|V|^2_t\,d\mu\,(t)\right\| \leq \int_E p\,(t)\,\big\||V|^2_t\big\|\,d\mu\,(t)$$

$$= \int_E p\,(t)\,\|V_t\|^2\,d\mu\,(t) = \big\|V_{(\cdot)}\big\|^2_{p,2}$$

and

$$\left\|\int_E p\,(t)\,|V^*|^2_t\,d\mu\,(t)\right\| \leq \int_E p\,(t)\,\big\||V^*|^2_t\big\|\,d\mu\,(t)$$

$$= \int_E p(t) \|V_t\|^2 d\mu(t) = \|V_{(\cdot)}\|_{p,2}^2,$$

then we have from (5.18) the following sequence of inequalities

$$\left\| \int_E p(t) V_t d\mu(t) \right\|^2 \tag{5.22}$$

$$\leq \|V_{(\cdot)}\|_{s,p,2}^2$$

$$\leq \left\| \int_E p(t) |V|_t^2 d\mu(t) \right\|^\alpha \left\| \int_E p(t) |V^*|_t^2 d\mu(t) \right\|^{1-\alpha}$$

$$\leq \alpha \left\| \int_E p(t) |V|_t^2 d\mu(t) \right\| + (1-\alpha) \left\| \int_E p(t) |V^*|_t^2 d\mu(t) \right\|$$

$$\leq \|V_{(\cdot)}\|_{p,2}^2$$

for any $V_{(\cdot)} \in \mathcal{L}_{2,p}(E, \mathcal{B}(H), \mu)$ and $\alpha \in [0, 1]$.
    From (5.19) we also have

$$w^2 \left(\int_E p(t) V_t d\mu(t) \right) \tag{5.23}$$

$$\leq w_{s,p,2}^2 (V_{(\cdot)})$$

$$\leq \begin{cases} \left\| \int_E p(t) |V|_t^2 d\mu(t) \right\|^\alpha \left\| \int_E p(t) |V^*|_t^2 d\mu(t) \right\|^{1-\alpha} \\ \left\| \int_E p(t) \left[\alpha |V|_t^2 + (1-\alpha) |V^*|_t^2 \right] d\mu(t) \right\| \end{cases}$$

$$\leq \alpha \left\| \int_E p(t) |V|_t^2 d\mu(t) \right\| + (1-\alpha) \left\| \int_E p(t) |V^*|_t^2 d\mu(t) \right\| \leq \|V_{(\cdot)}\|_{p,2}^2$$

for any $V_{(\cdot)} \in \mathcal{L}_{2,p}(E, \mathcal{B}(H), \mu)$ and $\alpha \in [0, 1]$.

    Now, we can consider the following Banach space $\mathcal{L}_{1,p}(E, \mathcal{B}(H), \mu)$ of all functions $V_{(\cdot)} : E \to \mathcal{B}(H)$ that are μ-measurable on E and such that

$$\|V_{(\cdot)}\|_{p,1} := \int_E p(t) \|V_t\| d\mu(t) < \infty.$$

We can consider in this space the following numerical radius

$$w_{p,1}(V_{(\cdot)}) := \int_E p(t) w(V_t) d\mu(t)$$

and taking into account the inequality (5.16) we can state that

$$\frac{1}{2} \left\| V_{(\cdot)} \right\|_{p,1} \leq w_{p,1} \left(V_{(\cdot)} \right) \leq \left\| V_{(\cdot)} \right\|_{p,1} \tag{5.24}$$

for any $V_{(\cdot)} \in \mathcal{L}_{1,p} \left(E, \mathcal{B} \left(H \right), \mu \right)$.

We can define the s-1-p-semi-norm by

$$\left\| V_{(\cdot)} \right\|_{s,p,1} := \sup_{\|x\|=\|y\|=1} \left(\int_E p \left(t \right) \left| \langle V_t x, y \rangle \right| d\mu \left(t \right) \right)$$

and the s-1-p-semi-numerical radius by

$$w_{s,p,1} \left(V_{(\cdot)} \right) := \sup_{\|x\|=1} \left(\int_E p \left(t \right) \left| \langle V_t x, x \rangle \right| d\mu \left(t \right) \right),$$

where $V_{(\cdot)} \in \mathcal{L}_{1,p} \left(E, \mathcal{B} \left(H \right), \mu \right)$.

Utilising the supremum properties we also have

$$\left\| V_{(\cdot)} \right\|_{s,p,1} \leq \left\| V_{(\cdot)} \right\|_{p,1} \text{ and } w_{s,p,1} \left(V_{(\cdot)} \right) \leq w_{p,1} \left(V_{(\cdot)} \right) \tag{5.25}$$

for any $V_{(\cdot)} \in \mathcal{L}_{1,p} \left(E, \mathcal{B} \left(H \right), \mu \right)$.

More related results are incorporated in the following theorem.

Theorem 5.8 (Dragomir [5]) *If* $|V|_{(\cdot)}^{2\alpha}$ *and* $|V^*|_{(\cdot)}^{2(1-\alpha)}$ *belong to* $\mathcal{L}_{1,p} \left(E, \mathcal{B} \left(H \right), \mu \right)$ *for some* $\alpha \in [0, 1]$, *then we have*

$$\left\| \int_E p \left(t \right) V_t d\mu \left(t \right) \right\| \tag{5.26}$$

$$\leq \left\| V_{(\cdot)} \right\|_{s,p,1}$$

$$\leq \left\| \int_E p \left(t \right) \left| V \right|_t^{2\alpha} d\mu \left(t \right) \right\|^{1/2} \left\| \int_E p \left(t \right) \left| V^* \right|_t^{2(1-\alpha)} d\mu \left(t \right) \right\|^{1/2}$$

and

$$w \left(\int_E p \left(t \right) V_t d\mu \left(t \right) \right) \tag{5.27}$$

$$\leq w_{s,p,1} \left(V_{(\cdot)} \right)$$

$$\leq \begin{cases} \left\| \int_E p\left(t\right)\left|V\right|_t^{2\alpha} d\mu\left(t\right) \right\|^{1/2} \left\| \int_E p\left(t\right)\left|V^*\right|_t^{2(1-\alpha)} d\mu\left(t\right) \right\|^{1/2}, \\ \frac{1}{2}\left\| \int_E p\left(t\right)\left[\left|V\right|_t^{2\alpha} + \left|V^*\right|_t^{2(1-\alpha)}\right] d\mu\left(t\right) \right\|. \end{cases}$$

Proof By the modulus properties and the inequality (5.10) we have

$$\left| \left\langle \left(\int_E p\left(t\right) V_t d\mu\left(t\right) \right) x, y \right\rangle \right|$$

$$\leq \int_E p\left(t\right) \left| \langle V_t x, y \rangle \right| d\mu\left(t\right)$$

$$\leq \left\langle \left(\int_E p\left(t\right) \left|V\right|_t^{2\alpha} d\mu\left(t\right) \right) x, x \right\rangle^{1/2} \left\langle \left(\int_E p\left(t\right) \left|V^*\right|_t^{2(1-\alpha)} d\mu\left(t\right) \right) y, y \right\rangle^{1/2},$$

for any $x, y \in H$.

Taking the supremum over $\|x\| = \|y\| = 1$ we deduce (5.26).

The second inequality follows by (5.10) and the details are omitted. ∎

Remark 5.9 Since by the integral triangle inequality for the norm we have

$$\left\| \int_E p\left(t\right) \left|V\right|_t^{2\alpha} d\mu\left(t\right) \right\| \leq \int_E p\left(t\right) \left\| \left|V\right|_t^{2\alpha} \right\| d\mu\left(t\right) = \int_E p\left(t\right) \left\|V_t\right\|^{2\alpha} d\mu\left(t\right)$$

and

$$\left\| \int_E p\left(t\right) \left|V^*\right|_t^{2(1-\alpha)} d\mu\left(t\right) \right\| \leq \int_E p\left(t\right) \left\| \left|V^*\right|_t^{2(1-\alpha)} \right\| d\mu\left(t\right)$$

$$= \int_E p\left(t\right) \left\|V_t^*\right\|^{2(1-\alpha)} d\mu\left(t\right)$$

then by (5.26) we have the following sequence of inequalities

$$\left\| \int_E p\left(t\right) V_t d\mu\left(t\right) \right\| \tag{5.28}$$

$$\leq \left\| V_{(\cdot)} \right\|_{s,p,1}$$

$$\leq \left\| \int_E p\left(t\right) \left|V\right|_t^{2\alpha} d\mu\left(t\right) \right\|^{1/2} \left\| \int_E p\left(t\right) \left|V^*\right|_t^{2(1-\alpha)} d\mu\left(t\right) \right\|^{1/2}$$

$$\leq \left(\int_E p\left(t\right) \left\|V_t\right\|^{2\alpha} d\mu\left(t\right) \right)^{1/2} \left(\int_E p\left(t\right) \left\|V_t\right\|^{2(1-\alpha)} d\mu\left(t\right) \right)^{1/2}$$

$$\leq \frac{1}{2} \int_E p\left(t\right) \left[\left\|V_t\right\|^{2\alpha} + \left\|V_t\right\|^{2(1-\alpha)} \right] d\mu\left(t\right)$$

provided that $|V|_{(\cdot)}^{2\alpha}$ and $|V^*|_{(\cdot)}^{2(1-\alpha)}$ belong to $\mathcal{L}_{1,p}\left(E, \mathcal{B}\left(H\right), \mu\right)$ for some $\alpha \in [0, 1]$.
    Under the same assumptions we also have

$$w\left(\int_E p\left(t\right) V_t d\mu\left(t\right)\right) \tag{5.29}$$

$$\leq w_{s,p,1}\left(V_{(\cdot)}\right)$$

$$\leq \begin{cases} \left\|\int_E p\left(t\right) |V|_t^{2\alpha} d\mu\left(t\right)\right\|^{1/2} \left\|\int_E p\left(t\right) |V^*|_t^{2(1-\alpha)} d\mu\left(t\right)\right\|^{1/2} \\ \frac{1}{2}\left\|\int_E p\left(t\right) \left[|V|_t^{2\alpha} + |V^*|_t^{2(1-\alpha)}\right] d\mu\left(t\right)\right\| \end{cases}$$

$$\leq \begin{cases} \left(\int_E p\left(t\right) \|V_t\|^{2\alpha} d\mu\left(t\right)\right)^{1/2} \left(\int_E p\left(t\right) \|V_t\|^{2(1-\alpha)} d\mu\left(t\right)\right)^{1/2} \\ \frac{1}{2}\int_E p\left(t\right) \left\|\left[|V|_t^{2\alpha} + |V^*|_t^{2(1-\alpha)}\right]\right\| d\mu\left(t\right) \end{cases}$$

$$\leq \frac{1}{2}\int_E p\left(t\right) \left[\|V_t\|^{2\alpha} + \|V_t\|^{2(1-\alpha)}\right] d\mu\left(t\right).$$

Remark 5.10   The case $\alpha = \frac{1}{2}$ is of interest since it generates from (5.29) the following inequalities

$$w_{s,p,1}\left(V_{(\cdot)}\right) \tag{5.30}$$

$$\leq \begin{cases} \left\|\int_E p\left(t\right) |V|_t d\mu\left(t\right)\right\|^{1/2} \left\|\int_E p\left(t\right) |V^*|_t d\mu\left(t\right)\right\|^{1/2} \\ \frac{1}{2}\left\|\int_E p\left(t\right) \left[|V|_t + |V^*|_t\right] d\mu\left(t\right)\right\| \end{cases}$$

$$\leq \begin{cases} \left\|V_{(\cdot)}\right\|_{p,1} \\ \frac{1}{2}\int_E p\left(t\right) \left\|\left[|V|_t + |V^*|_t\right]\right\| d\mu\left(t\right) \end{cases} \quad \left(\leq \left\|V_{(\cdot)}\right\|_{p,1}\right)$$

for any $V_{(\cdot)} \in \mathcal{L}_{1,p}\left(E, \mathcal{B}\left(H\right), \mu\right)$.

From (5.26), we also have for $\alpha = \frac{1}{2}$ the following refinement of the integral triangle inequality for norm:

$$\left\| \int_E p(t) V_t d\mu(t) \right\| \tag{5.31}$$

$$\leq \|V_{(\cdot)}\|_{s,p,1}$$

$$\leq \left\| \int_E p(t) |V|_t d\mu(t) \right\|^{1/2} \left\| \int_E p(t) |V^*|_t d\mu(t) \right\|^{1/2}$$

$$\leq \frac{1}{2} \left[\left\| \int_E p(t) |V|_t d\mu(t) \right\| + \left\| \int_E p(t) |V^*|_t d\mu(t) \right\| \right] \leq \|V_{(\cdot)}\|_{p,1}$$

for any $V_{(\cdot)} \in \mathcal{L}_{1,p}(E, \mathcal{B}(H), \mu)$.

Moreover, if the values of the function $V_{(\cdot)}$ are normal operators for μ-a.e. $t \in E$, then the inequality (5.31) becomes

$$\left\| \int_E p(t) V_t d\mu(t) \right\| \leq \|V_{(\cdot)}\|_{s,p,1} \leq \left\| \int_E p(t) |V|_t d\mu(t) \right\| \leq \|V_{(\cdot)}\|_{p,1}. \tag{5.32}$$

5.4 Applications for the Operator Exponential

It is known that if U and V are commuting operators, then the operator *exponential function* $\exp : \mathcal{B}(H) \to \mathcal{B}(H)$ given by

$$\exp(T) := \sum_{n=0}^{\infty} \frac{1}{n!} T^n$$

satisfies the property

$$\exp(U) \exp(V) = \exp(V) \exp(U) = \exp(U + V).$$

Also, if A is invertible and $a, b \in \mathbb{R}$ with $a < b$ then

$$\int_a^b \exp(tA) \, dt = A^{-1} \left[\exp(bA) - \exp(aA) \right].$$

We observe that if the values of the function $V_{(\cdot)}$ are normal operators for μ-a.e. $t \in E$, i.e., $|V|_t^2 = |V^*|_t^2$ for μ-a.e. $t \in E$, then we have from (5.7) that

$$\left| \left\langle \left(\int_E p(t) V_t d\mu(t) \right) x, y \right\rangle \right|^2 \tag{5.33}$$

$$\leq \left\langle \left(\int_E p(t) |V|_t^2 d\mu(t) \right) x, x \right\rangle^\alpha \left\langle \left(\int_E p(t) |V|_t^2 d\mu(t) \right) y, y \right\rangle^{1-\alpha}$$

for any $x, y \in H$ with $\|x\| = \|y\| = 1$ and $\alpha \in [0, 1]$.

If N is a normal operator, then for any real number t we have

$$|\exp(tN)|^2 = (\exp(tN))^* (\exp(tN)) = \exp(tN^*)(\exp(tN)) = \exp[t(N^* + N)].$$

Proposition 5.11 (Dragomir [5]) *Let N be an invertible normal operator and such that $N^* + N$ is also invertible. Then for any $a, b \in \mathbb{R}$ with $a < b$ we have*

$$\left| \left\langle N^{-1} \left[\exp(bN) - \exp(aN) \right] x, y \right\rangle \right|^2 \tag{5.34}$$

$$\leq \left\langle (N^* + N)^{-1} \left[\exp(b(N^* + N)) - \exp(a(N^* + N)) \right] x, x \right\rangle^\alpha$$

$$\times \left\langle (N^* + N)^{-1} \left[\exp(b(N^* + N)) - \exp(a(N^* + N)) \right] y, y \right\rangle^{1-\alpha}$$

for any $x, y \in H$ with $\|x\| = \|y\| = 1$ and $\alpha \in [0, 1]$.

Proof Follows from (5.33) applied for $E = [a, b]$, $p(t) = \frac{1}{b-a}$, $V_t = \exp(tN)$ and $d\mu(t) = dt$, the usual Lebesgue measure. ∎

If S is an invertible selfadjoint operator, then from (5.34) we get for any $a, b \in \mathbb{R}$ with $a < b$ that

$$\left| \left\langle S^{-1} \left[\exp(bS) - \exp(aS) \right] x, y \right\rangle \right|^2 \tag{5.35}$$

$$\leq \frac{1}{2} \left\langle S^{-1} \left[\exp(2bS) - \exp(2aS) \right] x, x \right\rangle^\alpha$$

$$\times \left\langle S^{-1} \left[\exp(2bS) - \exp(2aS) \right] y, y \right\rangle^{1-\alpha}$$

for any $x, y \in H$ with $\|x\| = \|y\| = 1$ and $\alpha \in [0, 1]$.

In particular, we have from (5.35) that

$$\left| \left\langle S^{-1} \left[\exp(S) - I \right] x, y \right\rangle \right|^2 \tag{5.36}$$

$$\leq \frac{1}{2} \left\langle S^{-1} \left[\exp(2S) - I \right] x, x \right\rangle^\alpha \left\langle S^{-1} \left[\exp(2S) - I \right] y, y \right\rangle^{1-\alpha}$$

for any $x, y \in H$ with $\|x\| = \|y\| = 1$ and $\alpha \in [0, 1]$.

If we use the inequality (5.10) for functions $V_{(\cdot)}$ that are normal operators for μ-a.e. $t \in E$, then we have

$$\left| \left(\left(\int_E p\,(t)\,V_t d\mu\,(t) \right) x, y \right) \right| \tag{5.37}$$

$$\leq \left(\left(\int_E p\,(t)\,|V|_t^{2\alpha}\,d\mu\,(t) \right) x, x \right)^{1/2} \left(\left(\int_E p\,(t)\,|V|_t^{2(1-\alpha)}\,d\mu\,(t) \right) y, y \right)^{1/2}$$

for any $x, y \in H$ and $\alpha \in [0, 1]$.

Utilising the inequality (5.37) we can state:

Proposition 5.12 (Dragomir [5]) *With the assumptions of Proposition 5.11, we have*

$$\left| \left\langle N^{-1} \left[\exp\,(bN) - \exp\,(aN) \right] x, y \right\rangle \right| \tag{5.38}$$

$$\leq \left\langle \alpha^{-1} \left(N^* + N \right)^{-1} \left[\exp\,(b\alpha\,(N^* + N)) - \exp\,(a\alpha\,(N^* + N)) \right] x, x \right\rangle^{1/2}$$

$$\times \left\langle (1-\alpha)^{-1} \left(N^* + N \right)^{-1} \right.$$

$$\left. \times \left[\exp\,(b\,(1-\alpha)\,(N^* + N)) - \exp\,(a\,(1-\alpha)\,(N^* + N)) \right] y, y \right\rangle^{1/2}$$

for any $x, y \in H$ *and* $\alpha \in (0, 1)$.

If S is an invertible selfadjoint operator, then from (5.38) we get for any $a, b \in \mathbb{R}$ with $a < b$ that

$$\left| \left\langle S^{-1} \left[\exp\,(bS) - \exp\,(aS) \right] x, y \right\rangle \right| \tag{5.39}$$

$$\leq \frac{1}{2\sqrt{\alpha\,(1-\alpha)}} \left\langle S^{-1} \left[\exp\,(2b\alpha S) - \exp\,(2a\alpha S) \right] x, x \right\rangle^{1/2}$$

$$\times \left\langle S^{-1} \left[\exp\,(2b\,(1-\alpha)\,S) - \exp\,(2a\,(1-\alpha)\,S) \right] y, y \right\rangle^{1/2}$$

for any $x, y \in H$ and $\alpha \in (0, 1)$.

In particular, we have from (5.39) that

$$\left| \left\langle S^{-1} \left[\exp\,(S) - I \right] x, y \right\rangle \right| \tag{5.40}$$

$$\leq \frac{1}{2\sqrt{\alpha\,(1-\alpha)}} \left\langle S^{-1} \left[\exp\,(2\alpha S) - I \right] x, x \right\rangle^{1/2}$$

$$\times \left\langle S^{-1} \left[\exp\,(2\,(1-\alpha)\,S) - I \right] y, y \right\rangle^{1/2}$$

for any $x, y \in H$ and $\alpha \in (0, 1)$.

The interested reader may state the corresponding norm inequalities. However the details are omitted.

References

1. W. Arveson, *A Short Course on Spectral Theory* (Springer Inc., New York, 2002)
2. S.S. Dragomir, The hypo-Euclidean norm of an n-tuple of vectors in inner product spaces and applications. J. Inequal. Pure Appl. Math. **8**(2), Article 52, 22 (2007)
3. S.S. Dragomir, Some inequalities of Kato type for sequences of operators in Hilbert spaces. Publ. RIMS Kyoto Univ. **46**, 937–955 (2012)
4. S.S. Dragomir, Generalizations of Furuta's inequality. Linear Multilinear Algebr. **61**(5), 617–626 (2013)
5. S.S. Dragomir, Applications of Kato's inequality to operator-valued integrals on Hilbert spaces. Asian-Eur. J. Math. **6**(4), 1350059 (18 pages) (2013)
6. S.S. Dragomir, Some inequalities of Furuta's type for functions of operators defined by power series. Acta Univ. Sapientiae Math. **6**(2), 162–177 (2014)
7. S.S. Dragomir, Some inequalities generalizing Kato's and Furuta's results. Filomat **28**(1), 179–195 (2014)
8. S.S. Dragomir, Some inequalities for trace class operators via a Kato's result. Asian-Eur. J. Math. **11**(1) 1850004 (24 pages) (2018)
9. S.S. Dragomir, Y.J. Cho, Y.-H. Kim, Applications of Kato's inequality for n-tuples of operators in Hilbert spaces, (I). J. Inequal. Appl. **2013**(21), 16 pp (2013)
10. S.S. Dragomir, Y.J. Cho, Y.-H. Kim, Applications of Kato's inequality for n-tuples of operators in Hilbert spaces, (II). J. Inequal. Appl. **2013**, 464, 20 pp (2013)
11. M. Fujii, C.-S. Lin, R. Nakamoto, Alternative extensions of Heinz-Kato-Furuta inequality. Sci. Math. **2**(2), 215–221 (1999)
12. M. Fujii, T. Furuta, Löwner-Heinz, Cordes and Heinz-Kato inequalities. Math. Japon. **38**(1), 73–78 (1993)
13. M. Fujii, E. Kamei, C. Kotari, H. Yamada, Furuta's determinant type generalizations of Heinz-Kato inequality. Math. Japon. **40**(2), 259–267 (1994)
14. M. Fujii, Y.O. Kim, Y. Seo, Further extensions of Wielandt type Heinz-Kato-Furuta inequalities via Furuta inequality. Arch. Inequal. Appl. **1**(2), 275–283 (2003)
15. M. Fujii, Y.O. Kim, M. Tominaga, Extensions of the Heinz-Kato-Furuta inequality by using operator monotone functions. Far East J. Math. Sci. (FJMS) **6**(3), 225–238 (2002)
16. M. Fujii, R. Nakamoto, Extensions of Heinz-Kato-Furuta inequality. II. J. Inequal. Appl. **3**(3), 293–302 (1999)
17. M. Fujii, R. Nakamoto, Extensions of Heinz-Kato-Furuta inequality. Proc. Am. Math. Soc. **128**(1), 223–228 (2000)

© The Author(s), under exclusive license to Springer Nature Switzerland AG 2019
S. S. Dragomir, *Kato's Type Inequalities for Bounded Linear Operators in Hilbert Spaces*, SpringerBriefs in Mathematics,
https://doi.org/10.1007/978-3-030-17459-0

18. T. Furuta, Determinant type generalizations of Heinz-Kato theorem via Furuta inequality. Proc. Am. Math. Soc. **120**(1), 223–231 (1994)
19. T. Furuta, An extension of the Heinz-Kato theorem. Proc. Am. Math. Soc. **120**(3), 785–787 (1994)
20. T. Furuta, Equivalence relations among Reid, Löwner-Heinz and Heinz-Kato inequalities, and extensions of these inequalities. Integr. Equ. Oper. Theory **29**(1), 1–9 (1997)
21. G. Helmberg, *Introduction to Spectral Theory in Hilbert Space* (Wiley, Inc., New York, 1969)
22. T. Kato, Notes on some inequalities for linear operators. Math. Ann. **125**, 208–212 (1952)
23. F. Kittaneh, Norm inequalities for fractional powers of positive operators. Lett. Math. Phys. **27**(4), 279–285 (1993)
24. F. Kittaneh, Notes on some inequalities for Hilbert space operators. Publ. Res. Inst. Math. Sci. **24**(2), 283–293 (1988)
25. C.-S. Lin, On inequalities of Heinz and Kato, and Furuta for linear operators. Math. Japon. **50**(3), 463–468 (1999)
26. C.-S. Lin, On Heinz-Kato type characterizations of the Furuta inequality. II. Math. Inequal. Appl. **2**(2), 283–287 (1999)
27. C.-S. Lin, On chaotic order and generalized Heinz-Kato-Furuta-type inequality. Int. Math. Forum **2**(37–40), 1849–1858 (2007)
28. C.-S. Lin, On Heinz-Kato-Furuta inequality with best bounds. J. Korea Soc. Math. Educ. Ser. B Pure Appl. Math. **15**(1), 93–101 (2008)
29. C.A. McCarthy, c_p. Israel J. Math. **5**, 249–271 (1967)
30. G. Popescu, Unitary invariants in multivariable operator theory. Mem. Am. Math. Soc. **200**(941), vi+91 (2009)
31. B. Simon, *Trace Ideals and Their Applications* (Cambridge University Press, Cambridge, 1979)
32. V.I. Sobolev, Bochner integral. Encycl. Math. http://www.encyclopediaofmath.org/index.php
33. M. Uchiyama, Further extension of Heinz-Kato-Furuta inequality. Proc. Am. Math. Soc. **127**(10), 2899–2904 (1999)

Printed in the United States
By Bookmasters